Oxford Physics Series

General Editors

E. J. BURGE   **D. J. E. INGRAM**   **J. A. D. MATTHEW**

Oxford Physics Series

1. F.N.H. Robinson: *Electromagnetism*

2. G. Lancaster: *D.c. and a.c. circuits*

3. D.J.E. Ingram: *Radiation and quantum physics*

4. J.A.R. Griffith: *Interactions of particles*

5. B.R. Jennings and V.J. Morris: *Atoms in contact*

6. G.K.T. Conn: *Atoms and their structure*

7. R.L.F. Boyd: *Space physics; the study of plasmas in space*

8. J.L. Martin: *Basic quantum mechanics*

9. H.M. Rosenburg: *The solid state*

10. J.G. Taylor: *Special relativity*

11. M. Prutton: *Surface physics*

12. G.A. Jones: *The properties of nuclei*

13. E.J. Burge: *Atomic nuclei and their particles*

14. W.T. Welford: *Optics*

15. M. Rowan-Robinson: *Cosmology*

16. D.A. Fraser: *The physics of semiconductor devices*

MICHAEL ROWAN-ROBINSON
Queen Mary College, London

Cosmology

Clarendon Press.   Oxford

Oxford University Press, Walton Street, Oxford OX2 6DP

OXFORD LONDON GLASGOW NEW YORK
TORONTO MELBOURNE WELLINGTON CAPE TOWN
IBADAN NAIROBI DAR ES SALAAM
KUALA LUMPUR SINGAPORE JAKARTA HONG KONG TOKYO
DELHI BOMBAY CALCUTTA MADRAS KARACHI

First published 1977
Reprinted 1978

British Library Cataloguing in Publication Data

Rowan-Robinson, Michael
  Cosmology. - (Oxford physics series; no.15).
  1. Cosmology
  I. Title II. Series
  523.1   QB981

  ISBN 0-19-851838-2
  ISBN 0-19-851839-0 Pbk

Printed in Great Britain
by J.W.Arrowsmith Ltd., Bristol

Preface

From the earliest times man has asked questions about the
universe he finds himself in. We are fortunate to be alive
in the third great age of cosmology. The first was the age
of Epicurus, Aristarchus, and Hipparchus, the third and second
centuries B.C., when the notion of an infinite universe in which
the earth is not at the centre was first considered, though not
adopted. The second began with the methodical programme of
Copernicus to prove the motion of the earth and ended with that
great cry of Bruno: 'The stars are suns like our own and there
are countless suns freely suspended in limitless space, all of
them surrounded by planets like our own earth, peopled with
living beings. The sun is only one star among many, singled out
because it is so close to us. The sun has no central position
in the boundless infinite.'

In the third age of cosmology, which could be said to have
begun with Einstein's proposal of an isotropic and homogeneous
universe, we are seeing the whole electromagnetic spectrum, from
radio through microwave and infrared to X- and γ-rays, pressed
into the service of cosmology. We are probably only at the
beginning of the revolution that these new windows on the
universe will bring about.

Despite the wealth of new information, the models of the
universe that are now favoured are the simplest big-bang models
put forward by Einstein, de Sitter, Friedmann, and Lemaître in
the 1920s. The decisive factor in this convergence of thinking
has been the cosmic microwave radiation discovered by Penzias
and Wilson in 1965. For no explanation of this other than that
it is the relic of the fireball phase of an isotropic big bang
has stood the test of time. The steady-state cosmology has

long since fallen by the wayside.

This therefore seems a good moment to write a book
outlining modern ideas about cosmology at an elementary level.
The book appears in a series aimed at first-year maths and
physics undergraduates, but I hope that most of it will be
accessible to those who have studied science to an advanced
level at school.

Because this is neither a research text nor a history of
cosmology I have abandoned the practice, alas seldom strictly
adhered to, of giving references to all sources of ideas and
facts. The reader will easily see the extent to which modern
science is a collective enterprise by consulting the review
articles listed at the end of this book, which between them refer
to the literally thousands of scientific papers on which the
picture presented here is based. I crave the indulgence of the
many hundreds of colleagues who should have been acknowledged
in this way. They at least will know the part they have played
in this book.

As I remain deeply sceptical of many of the ideas outlined
here, I have tried to emphasize the open-endedness of cosmology,
the fact that the whole argument is still going on. To
underline this I have included an epilogue outlining twenty
controversies in cosmology today.

I must record my thanks to Professor W.H.McCrea, both for
his lectures on cosmology which first aroused my interest in
the subject and for reading and commenting on the manuscript of
this book. My thanks also to several colleagues and friends
who read the manuscript, made suggestions, and corrected errors,
especially Wal Sargent, Laura Maraschi, Ian Roxburgh, Paul
Davies, and Andy Fabian, but they are in no way responsible for
the errors and deficiencies that remain. I owe a special debt
of gratitude to Professor R.F. Boyd for his encouragement and
interest in the book from its earliest stages and for his
careful reading and criticism of the first draft.

January 1975   M.R-R.

Acknowledgements

Photographs were kindly provided by A. Penzias (frontispiece),
B. Elsemore (Fig. 1.9 (a)), R. Sancisi (Figs. 1.9(b) and 2.7),
R.F. Boyd (Fig. 1.9(c)),K. Pounds (Figs. 1.9(d) and 2.12), and
R. Carswell (Fig. 1.11). Permission to use copyright material
was kindly given by: John Stewart (Fig. 1.9(a)); Luchthaven
Eelde and the Netherland Foundation for Radioastronomy (Fig.
1.9(b)); J.B. Whiteoak (Fig. 2.4(e)); P.C. van der Kruit (Fig.2.7);
P.J. Hargrave, M. Ryle and the Royal Astronomical Society
(Fig. 2.8); the University of Chicago Press (Fig. 2.11); C.D.
Shane and the Lund Observatory (Fig. 3.4). I am grateful to
Mrs. Lake at the Royal Astronomical Society for providing the
Hale Observatory photographs. Jill Hailey somehow managed to
decipher my writing and produce a typescript.

This world has persisted many a long year, having once been set going in the appropriate motions. From these everything else follows.

LUCRETIUS

Do you want to stride into the infinite?
Then explore the finite in all directions.

GOETHE

And I say to any man or woman. Let your souls stand cool and composed before a million universes.

WHITMAN

Contents

1. THE VISIBLE UNIVERSE   1

    1.1 Introduction. 1.2. The electromagnetic spectrum.
    1.3. Astronomy without light. 1.4. Radiation mechanisms.
    1.5. Observing techniques at different frequencies.
    1.6. The brightest sources. 1.7. Source counts.
    1.8. Integrated background radiation. Problems.

2. OUR GALAXY, AND OTHER GALAXIES   23

    2.1. Introduction. 2.2. Star formation. 2.3. The
    evolution of a star. 2.4. Final stages: white dwarfs,
    neutron stars, and black holes. 2.5. The life history of
    our Galaxy. 2.6. The structure and evolution of galaxies.
    2.7. Radio properties of galaxies. 2.8. 'Active' galaxies
    and quasars. 2.9. Clusters of galaxies. Problems.

3. THE EMPIRICAL BASIS FOR COSMOLOGICAL THEORIES   45

    3.1. Introduction. 3.2. The distance scale. 3.3. The
    red-shift. 3.4. Isotropy. 3.5. Uniformity. 3.6. Olbers'
    paradox. 3.7. Evidence for a universe of finite age.
    3.8. Evidence for a 'fireball' phase. Problems.

4. THE BIG-BANG MODELS   63

    4.1. The substratum and fundamental observers. 4.2. The
    cosmological principle. 4.3. Newtonian cosmology.
    4.4. The special and general theories of relativity.
    4.5. General relativistic cosmology. 4.6. Classification
    of cosmological models. 4.7. Cosmological parameters.
    4.8. The age of the universe. Problems.

5. EARLY STAGES OF THE BIG BANG   82

    5.1. Universe with matter and radiation. 5.2. The
    fireball. 5.3. Helium production. Problem.

6. OBSERVATIONAL COSMOLOGY   93

    6.1. Introduction. 6.2. Newtonian theory. 6.3. Special
    relativity cosmology: the Milne model. 6.4. General
    relativistic cosmology: the red-shift. 6.5. Luminosity
    distance. 6.6. Diameter distance. 6.7. Number counts of
    sources. 6.8. The luminosity-volume test. 6.9. Integrated
    background radiation. 6.10. Horizon. Problem.

7. THE DENSITY OF MATTER IN THE UNIVERSE   113

    7.1. Introduction. 7.2. Masses of galaxies. 7.3. The average density of matter in galaxies. 7.4. Clusters of galaxies. 7.5. Some other possible forms of matter. 7.6. Intergalactic gas. 7.7. Formation of galaxies. Problems.

8. OTHER COSMOLOGICAL THEORIES   123

    8.1. General relativistic models with the Λ-term. 8.2. Observational consequences of the Λ-term. 8.3. The steady-state cosmology. 8.4. Theories in which G changes with time. 8.5. Anisotropic universes in general relativity. 8.6. The hierarchical universe. 8.7. Eddington's magic numbers. Problem.

EPILOGUE   136

    Twenty controversies in cosmology today

FURTHER READING   146

GLOSSARY   147

NAME INDEX   153

SUBJECT INDEX   154

The horn antenna with which, in 1965, Penzias and Wilson discovered the cosmic microwave background radiation, believed to be a relic of the 'fireball' phase of a big-bang universe.

1. The visible universe

1.1. INTRODUCTION

Imagine that it is a clear moonless night in the country.
The sky is ablaze with thousands of stars. The light from
some of the faintest ones set out thousands of years ago.

In a great arc across the sky we see that familiar band
of light, the Milky Way, which Galileo showed to be composed
of myriads of faint stars. This is our own Galaxy, a discus-
shaped metropolis of a hundred thousand million (10^{11}) stars
with ourselves far out towards the suburbs (see Fig. 1.1)

FIG. 1.1. The outline of our Galaxy, as it would look edge-on
The small circle, centred on where the sun would be, indicates
the region where most of the stars visible to the naked eye lie.
The Milky Way consists of the integrated light from more distant
stars in the disc.

Less than 50 years ago it was still reasonable to believe
that this gigantic star system, 60,000 light years in diameter
(1 light year = distance travelled by light in one year =
0.946×10^{16}m) comprised the whole visible universe. Today our
horizon is at least 100,000 times larger.

As a first step out from our own Galaxy let us, with the
aid of a large telescope, travel towards a faint and fuzzy
patch of light in the constellation of Andromeda, the nebula
M31 (see Figs. 1.2 and 1.3). Two million light years away, it
is almost the twin of our own Galaxy, seen tilted to our line
of sight. These two galaxies are two of the dominant members
of a small group of 20 or so galaxies known as the Local Group
of galaxies (see Table 1; p.4).

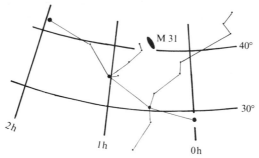

FIG. 1.2. The constellation of Andromeda, showing the location of the nebula, Messier 31. In 1924 Hubble showed that M31 lies far outside our Galaxy. The coordinates shown are Right Ascension (horizontal axis) and declination (vertical axis), which are equivalent to latitude and longitude projected onto the sky.

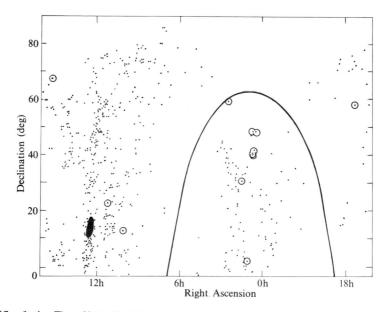

FIG. 1.4. The distribution of northern-hemisphere galaxies brighter than 13th magnitude on the sky. The solid line marks the plane of the Milky Way, near which very few galaxies are detectable due to obscuration by dust. The shaded region near $(12\,h, 15°)$ denotes the core of the Virgo cluster of galaxies, with too many galaxies to be shown individually. Circled points are Local-Group members.

FIG. 1.3. The Andromeda nebula, a spiral galaxy seen tilted to our line of sight. Messier 31 and our own Galaxy are the dominant members of the Local Group of galaxies. Two other members, the dwarf elliptical galaxies NGC205 and 221, can also be seen in the photograph, which is from Hale Observatories. (Copyright by the California Institute of Technology and the Carnegie Institute, Washington).

Now let us travel 75 million light years towards the constellation of Virgo. We find ourselves in a cloud of thousands of galaxies, the Virgo cluster, first recognized by William Herschel (see Fig. 1.4). Our Galaxy may lie in the outer fringes of this cluster.

Suppose we travel out to the limit of vision of the 5m telescope on Mount Palomar, U.S.A. The most distant galaxies we can see are at least 10^{10} light years away. Their light

Table 1
The local group of galaxies

Name	Type (see p. 34)	Distance (light years)	lg (mass of galaxy/mass of sun, $M/M_\odot$)	Absolute visual magnitude (see p. 46)	Linear diameter (10^3 light years)	Radial velocity (km s^{-1})
M31	Sb	2200	11·5	−21·1	50	−275
Our Galaxy	Sab?	33 (to centre)	11·2	−20·5	80	0
M33 (NGC598)	Sc	2400	10·1	−18·8	20	−190
Large Magellanic cloud	Irr	170	10·0	−18·7	24	270
NGC205	E5	2100	9·9	−16·3	6	−240
M32 (NGC221)	E2	2200	9·5	−16·3	3	−210
Small Magellanic cloud	Irr	210	9·3	−16·7	10	168
NGC147	E$_{pec}$	2200	9	−14·8	3	−250
NGC185	E$_{pec}$	2200	9	−15·2	3	−300
NGC6822	Irr	1500	8·5	−15·6	6	−40
IC1613	Irr	2400	8·4	−14·8	3	−240
Fornax	E	550	7·3	−13	6	40
Leo I	E4	750	6·6	−11	3	
Sculptor	E	280	6·5	−12	3	
Leo II	E1	750	6·0	−9·5	3	
Draco	E	220	5	−8·5	1	
Ursa minor	E	220	5	−9	3	

set out long before the earth was formed. This whole expanse
is filled with galaxies and clusters of galaxies. This
'realm of the nebulae', as Hubble called it (see references,
p. 146) is the subject of this book.

1.2. THE ELECTROMAGNETIC SPECTRUM

Since even the nearest star, α Cen, 4 light years away, lies
for the moment far beyond man's reach, we can learn about
distant parts of the universe only from the light and other
kinds of information they send us.

The human eye responds to only a very narrow range of
frequencies, the *visible* portion of the electromagnetic spectrum
(Fig. 1.5). In 1800 Herschel first showed the possibility of

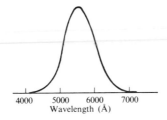

FIG. 1.5. The response function of the human eye, as a function
of wavelength.

using other frequencies or wavelengths for astronomy. He held
a thermometer beyond the red portion of the spectrum obtained
by passing the sun's light through a prism, demonstrating the
existence of *infrared* radiation from the sun (Fig. 1.6). Soon
afterwards Ritter found *ultraviolet* radiation. But it was not
until 1931 that the American radio amateur Jansky first showed
that the Milky Way emitted *radio* waves. And in 1948 a rocket
was launched carrying a camera that recorded *X-rays* from the
sun. The detectors used to record these different types of
radiation vary widely, from radio receivers and geiger counters

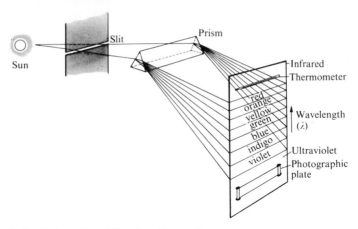

FIG. 1.6. Schematic illustration of the formation of a spectrum of the sun's light by means of a prism.

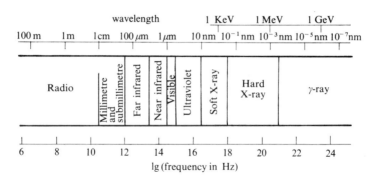

FIG. 1.7. The electromagnetic spectrum.

to bolometers and photographic plates. And because many types of radiation are absorbed by the atmosphere, a great variety of observing platforms are used, from high-altitude observatories to aeroplanes, balloons, and satellites.

But it is very important to be clear that these different radiations are all fundamentally the same, namely *electromagnetic radiation*, or more simply 'light'. The wavelength λ and frequency ν are related by the equation

$$\lambda\nu = c \qquad (1.1)$$

where c is the velocity of light. Fig. 1.7 illustrates how thee electromagnetic spectrum is arbitrarily divided into bands, defined primarily by the different detection techniques used. Radio, visible, and X-ray photons differ only in their frequency (and wavelength). Since a photon carries an energy $h\nu$ where h is Planck's constant, X-ray photons carry far more energy than, for example, radio photons. Table 2 (p.20) summarizes the electromagnetic spectrum for astronomy.

1.3. ASTRONOMY WITHOUT LIGHT

Electromagnetic radiation is not the only way that astronomical information reaches the earth.

Cosmic rays

These energetic particles, electrons and the nuclei of atoms, moving at velocities very close to the speed of light, bombard the solar system continuously from all directions. While you are reading this sentence one will probably pass through your head. The kinetic energies of some of these particles far exceed anything achieved in the largest man-made accelerator (Fig. 1.8). On entry to the atmosphere, the higher-energy nuclei soon collide with molecules of air, creating a shower of secondary particles, which can be detected on the ground.

The sun is an important source of low-energy cosmic rays. The main source of the more energetic particles is unknown, but it is believed that pulsars (pulsating radio sources associated with neutron stars), supernovae (explosions in high-mass stars that have exhausted their nuclear fuel), active galaxies (in which violent and explosive events of some kind are taking place), and quasars (quasi-stellar radio sources) all contribute.

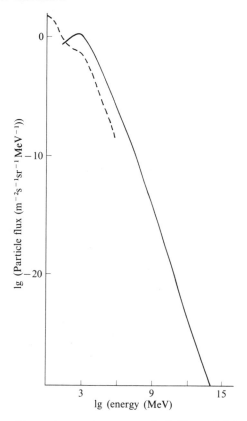

FIG. 1.8. The flux of cosmic rays of different kinetic energies reaching the top of the earth's atmosphere. Solid line: protons and nuclei. Broken line: electrons.

Neutrinos

These massless, chargeless, particles were first postulated by Pauli and Fermi to explain an energy imbalance in β-radioactivity. Like photons, they move at the speed of light. They are exceedingly hard to detect, since the probability of their taking part in any nuclear reaction under terrestial conditions is very low (they have a very low 'cross-section' for collisions with matter).

FIG. 1.9. Some of the more exotic telescopes of modern astronomy:
(a) the University of Cambridge - Science Research Council 5-km
radio telescope; (b) the Westerborg 3-km radio-telescope
(courtesy of the Netherlands Foundation for Radioastronomy);
(c) University College, London, X-ray telescope now on board the
Copernicus satellite; (d) the X-ray satellite UK5 prior to launch:
the lower dark square is the University of Leicester sky survey
experiment; (e) one of Weber's gravitational wave detectors.

(e)

R. Davis has created the world's first neutrino 'telescope' in South Dakota, U.S.A. It consists of a vast tank of perchloroethylene (C_2Cl_4), at the bottom of a gold mine to avoid contamination by cosmic rays. Any neutrinos passing through it have a small chance of converting a chlorine atom ^{37}Cl to an argon atom ^{37}Ar. These argon atoms are then 'counted'. Unfortunately Davis has not so far detected any cosmic neutrinos, although according to the original predictions of experts on the internal structure of the sun he should have done so by now.

Gravitational radiation

According to the general theory of relativity, and other similar theories of gravitation, gravitational waves should exist. The final stages of gravitational collapse of massive stars provide the most likely source of gravitational radiation. Weber has attempted to detect such radiation by looking for vibrations in a large, freely suspended aluminium bar (see Fig. 1.9). Although he did report a surprisingly large number of gravitational-wave events from the direction of the centre of our Galaxy, this has not so far been confirmed by other workers.

1.4. RADIATION MECHANISMS

We now look at the main ways in which electromagnetic radiation is produced in cosmic sources. We first define the *monochromatic luminosity* $P(\nu)$ of a source at frequency ν as the energy it emits per unit solid angle per unit frequency (units: $W\ sr^{-1}\ Hz^{-1}$). This means that the energy emitted by the source per steradian between frequencies ν and $\nu + d\nu$ is $P(\nu)\ d\nu$, and the total ('bolometric') luminosity can be obtained by integrating over all frequencies

$$P = \int_0^\infty P(\nu)\ d\nu.$$

The function $P(\nu)$, as a function of ν, is the spectrum or energy distribution of the source, and is an important clue to the radiation mechanism.

Black-body radiation

If a piece of matter completely absorbs all the radiation falling upon it or, conversely, behaves as a perfect radiator when heated, then the matter radiates as a blackbody, and the radiation has the characteristic Planck black-body spectrum: an element of the surface radiates with intensity

$$I(\nu) = 2h\nu^3/c^2 \{\exp(h\nu/kT) - 1\}\ W\ m^{-2}\ Hz^{-1}\ sr^{-1}, \tag{1.2}$$

where h is Planck's constant, k is the Boltzmann constant, and

T is the temperature. For

$$h\nu \ll kT, \exp(h\nu/kT) - 1 \simeq h\nu/kT, \text{ so } I(\nu) \propto \nu^2.$$

This is called the Rayleigh-Jeans part of the spectrum. For

$$h\nu \gg kT, I(\nu) \propto \nu^3 \exp(-h\nu/kT),$$

the Wien distribution.

The spectrum is shown in Fig. 1.10: the peak occurs at frequency $\nu = 0.354 \, kT/h$. Now kT is a measure of the average thermal energy

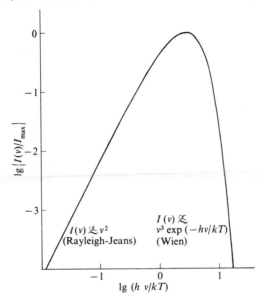

FIG. 1.10. The Planck black-body spectrum. The intensity is given in units of the peak intensity, $I_{max} = 1.9 \times 10^{-19} T^3$ $W \, m^{-2} \, sr^{-1} \, Hz^{-1}$.

of a particle of the matter (atom or molecule) and, from quantum theory, $h\nu$ is the energy of a photon with frequency ν. Thus most of the energy in black-body radiation is emitted at frequencies such that the energy of the photons is of the same order as the mean kinetic energy of the particles of matter. The visible light from the sun and stars corresponds approximately to a

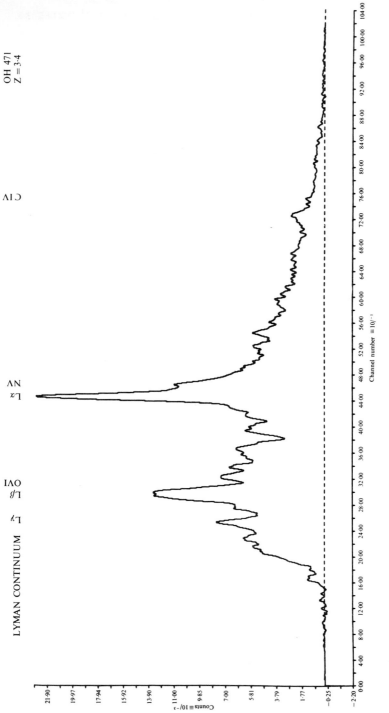

FIG. 1.11. The spectrum of the quasar OH471, showing both emission and absorption lines. The two strongest emission lines are the Lyman α and β lines of hydrogen. Spectrum obtained by J.B. Oke and A. Boksenburg.

blackbody. For the sun the temperature is 5800 K: most stars
have black-body temperatures in the range 2000-50 000 K.

Line radiation and absorption

When an electron in an atom makes a transition from an energy
level E_1 to a lower level E_2 a photon is emitted with frequency ν,
where $h\nu = E_1 - E_2$. Similarly an atom may absorb a photon of
energy $h\nu$ when an electron makes a transition from E_2 to the
higher energy level E_1. These processes are very important in
the surface layers of stars and result in clear emission and
absorption spikes in the energy distribution $P(\nu)$ (see Fig. 1.11).
If the light from a star is viewed through a prism and a narrow
slit, bright or dark lines appear across the spectrum. Such lines
were first noticed by Fraunhofer in the sun's spectrum. They
allow both the composition of, and the physical conditions in,
the surface layers of stars to be studied, since the frequencies
of the lines allow the emitting or absorbing atoms to be
identified and relative strengths of different lines give
information on temperature and the number of atoms involved.

One of the most important spectral lines for astronomy is the
21-cm radio line of neutral hydrogen, resulting from the transition from alignment to non-alignment and vice versa, of the
electron and proton magnetic axes. The distribution of neutral
hydrogen in the Milky Way and other galaxies has been mapped
using this line.

An important development in recent years has been the discovery
of line radiation from interstellar molecules, arising from
transitions between different states of rotational energy of the
molecules.

Synchrotron radiation

The radio emission from the Milky Way comes from cosmic-ray
electrons spiralling in the Galaxy's magnetic field (Fig. 1.12).
The process was first observed in man-made particle accelerators;
hence the name. It is believed to be the mechanism operating in

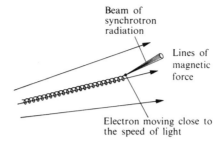

FIG. 1.12. A beam of synchrotron radiation from a relativistic
electron spiralling in a magnetic field.

the majority of cosmic radio sources. If the relativistic (i.e.
moving close to the speed of light) electrons have a power-law
distribution of energies (as we know they do in the vicinity of
the earth - See Fig. 1.8.), then the spectral energy distribution
of the synchrotron radiation will also be a power law (Fig. 1.13):

$$P_s(\nu) \propto (KB^{1 + \alpha})\nu^{\alpha}, \tag{1.3}$$

where B is the magnetic field intensity, K is the total energy
in relativistic electrons, and α is the spectral index.

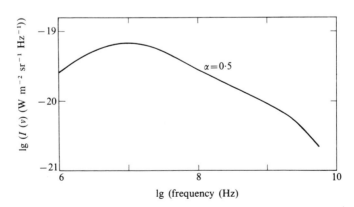

FIG. 1.13. The radio spectrum of the Milky Way. Over a wide range
of frequencies it shows the power-law behaviour expected if the
emission is synchrotron radiation due to cosmic-ray electrons
(Fig. 1.8) spiralling in our Galaxy's magnetic field.

Inverse Compton radiation

This is the inverse process to Compton scattering of light by free electrons (see *Radiation and quantum physics* by D.J.E. Ingram (OPS 3)). Relativistic electrons transfer some of their energy by collisions to photons in a radiation field, the photons emerging at a higher frequency (and energy). For example, radio photons might be boosted to become X-ray photons. The main requirement is that the initial radiation field have a high intensity, so this process is likely to be important in compact sources like quasars and galactic nuclei. In fact a relativistic electron will lose its energy by synchrotron radiation in a magnetic field or by inverse Compton radiation in a radiation field, when both are present, in the proportion of the respective energy densities of the magnetic and radiation fields.

1.5. OBSERVING TECHNIQUES AT DIFFERENT FREQUENCIES

Fortunately for life on earth, only certain of the radiations entering the top of the atmosphere reach the ground. Water vapour and other molecules absorb most radiation with wavelengths between $1\mu m$ (10^{-6} m) and 2 mm, although there are a few 'windows' (narrow ranges of wavelengths through which an appreciable part of the light is transmitted by the atmosphere). The atmosphere also transmits almost no radiation with wavelength less than 3000 Å ($0.3\mu m$). Anyway, far ultraviolet and soft (i.e. low-frequency) X-radiation can travel only comparatively short distances through the interstellar material (gas and dust) in our Galaxy, so we can not see very distant objects at these wavelengths even from outside the atmosphere. Low-frequency radio waves are totally reflected by the ionosphere (the same effect that makes intercontinental long-wave radio communications possible).

The percentage of light from outside our Galaxy reaching the surface of the earth is shown in Fig. 1.14 as a function of frequency. Even from a satellite we would not be able to

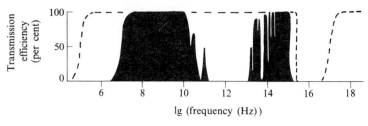

FIG. 1.14. Transmission efficiency of the earth's environment: the percentage of light from outside the Galaxy reaching the surface of the earth (solid curve) and a satellite outside the earth's atmosphere (broken curve), as a function of frequency.

receive some frequencies from outside our Galaxy. In addition to those mentioned above, our Galaxy is opaque to the very lowest energy radio waves and the very highest energy γ-rays.

With difficulty, sources radiating at wavelengths near 1 mm, 350 μm, 20 μm, and 10 μm, can be detected from high-altitude observatories. Balloons and aircraft have contributed to opening up these and other windows in the infrared and submillimetre region. Rockets and satellites have made ultraviolet and X-ray astronomy possible. Table 2 (p.20) summarizes the observing platforms and detectors for different ranges of frequency.

1.6. THE BRIGHTEST SOURCES

The *flux density* of radiation (energy per second per unit bandwidth per unit area normal to the direction of propagation) that we measure from a source of monochromatic luminosity $P(\nu)$ at distance r is given by the inverse square law

$$S(\nu) = P(\nu)/r^2 \quad (\text{W m}^{-2} \text{ Hz}^{-1}), \tag{1.4}$$

provided cosmological effects (e.g. the expansion of the universe), absorption, etc. can be neglected. Thus the brightest sources we observe may be either weaker emitters that are very

nearby or distant objects that are very luminous. In Table 2
the brightest sources in the different frequency bands are shown.

1.7. SOURCE COUNTS

Suppose that at frequency ν we catalogue all the sources in the
sky as a function of the flux density S (ν). Let N_ν (S) be the
number of sources per steradian having flux density greater than
S at frequency ν. We shall see later that this function provides
an important test of cosmological models. At this point we
merely note the faintest flux densities reached by present-day

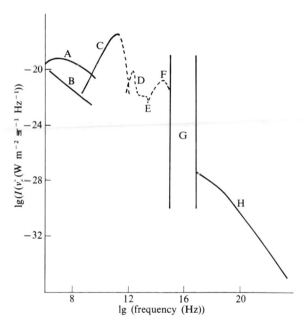

FIG. 1.15. Observed intensity of the integrated background
radiation as a function of frequency (broken sections indicating
theoretical predictions): A, background from the Milky Way
(Fig. 1.13); B, contribution of extragalactic sources with steep
radio spectra; C, 2.7°K black-body cosmic microwave background;
D,E,F, predicted contributions of emission from dust in our
Galaxy, from the 12.8 µm Neon line in galaxies, and from the combined
effects of atmospheric emission, zodaical light, and scattered
starlight; G, unobservable ultraviolet; H, the X- and γ-ray
background.

TABLE 2
The electromagnetic spectrum for astronomy

Band	Radio	Microwave	Submillimetre	Far infrared	Near infrared
Frequency range	10^6–10^{10} Hz	10^{10}–$10^{11.5}$ Hz	$10^{11.5}$–$10^{12.5}$ Hz	$10^{12.5}$–$10^{14.2}$ Hz	$10^{14.2}$–$10^{14.6}$ Hz
Wavelength range	3 cm–300 m	1 mm–3 cm	100 μm–1 mm	2–100 μm	0.8–2 μm
Observing platform	Ground (satellite for $\nu < 3\times10^7$ Hz)	Ground	Mountain (1 mm, 300 μm) balloon, rocket, aircraft (100, 300 μ)	Ground, aircraft, balloon	Ground
Detector	Radio receiver	Radio receiver	Bolometer	Bolometer	Photograph, image tube
Brightest observed sources	Cas A (supernova remnant) Sun Cyg A (radio galaxy) Sag A (galactic centre) Cen A (radio galaxy) Vir A (radio galaxy)	Sun Crab nebula (supernova remnant) Omega nebula Cas A Orion nebula Radio galaxies Quasars	Sun Planets Galactic centre Orion nebula Hydrogen and dust clouds in the Galaxy 3C273 (quasar) 3C84 (Seyfert galaxy) M82 (exploding galaxy)	Sun Planets Hydrogen and dust clouds in the Galaxy Stars in dust clouds Seyferts and 'active' galaxies Quasars	Sun Planets Cool stars or stars in dust clouds Galaxies Quasars
Faintest detectable flux (centre of band)	10^{-28} W m^{-2} Hz^{-1}	10^{-26} W m^{-2} Hz^{-1}	10^{-24} W m^{-2} Hz^{-1}	10^{-27} W m^{-2} Hz^{-1}	
Approximate number of sources in sky to this level	10^6	10^3	100	10^3	
Main contributing types of sources	Radio galaxies, quasars	Radio galaxies, quasars	Galactic sources	Stars, galaxies	
Main contribution to background radiation	Milky Way, radio galaxies	Cosmic, 2.7 K black body	Atmosphere	Atmosphere	Atmosphere, Zodiacal light

TABLE 2 (cont.)

Band	Visible	Ultraviolet	Soft X-ray	Hard X-ray	γ-ray
Frequency range	$10^{14\cdot6}-10^{14\cdot9}$ Hz	$10^{14\cdot9}-10^{16\cdot5}$ Hz	$10^{16\cdot5}-10^{17\cdot5}$ Hz	$10^{17\cdot5}-10^{20}$ Hz	10^{20} Hz
Wavelength range	0·4–0·8 μm	100–4000 Å	10–100 Å (0·12–1·2 keV)	(1·2–370 keV)	(> 370 keV)
Observing platform	Ground	Ground (3000–4000 Å) Rocket, satellite (λ < 3000 Å)	Rocket, satellite	Balloon, satellite	Balloon, satellite
Detector	Photograph, image tube	Photograph, image tube	Photon counters		
Brightest observed sources	Sun Planets Stars Galaxies Quasars	Sun Hot stars Galaxies Quasars	Sun Galactic sources (binary systems with white dwarf, neutron star, or black hole) Rich clusters of galaxies M82 (exploding galaxy) 3C273 (quasar) NGC4151 (Seyfert galaxy) M31 (galaxy) Magellanic clouds		Crab pulsar Galactic disc Vela pulsar
Faintest detectable flux (centre of band)	10^{-31} W m^{-2} Hz^{-1}	10^{-31} W m^{-2} Hz^{-1}	10^{-32} W m^{-2} Hz^{-1}		10^{-34} W m^{-2} Hz^{-1}
Approximate number of sources in sky to this level	10^9		100		3
Main contributing types of sources	Stars, galaxies		Galactic sources, clusters of galaxies		
Main contribution to background radiation	Atmosphere, zodiacal light, scattered starlight (galaxies)		Extragalactic sources ?		Extragalactic sources ?

techniques in the different frequency bands, the corresponding value of N_ν (S), and the main contributing types of source (Table 2).

1.8. INTEGRATED BACKGROUND RADIATION

If we point a telescope at a region of the sky free of bright sources, we can measure the total integrated background flux density from all the sources in the sky at all distances. This will depend on the size of the telescope beam, so we naturally measure this background in terms of the *intensity* of the radiation, the flux density per unit solid angle, I (ν) W m^{-2} Hz^{-1} sr^{-1}. The current state of observations of this quantity is shown in Fig.1.15. Only at radio and X-ray frequencies have we succeeded in detecting a background that probably does come from discrete sources. The most striking feature of Fig. 1.15. is the 2.7 K black-body background radiation at centimetre and millimetre, wavelengths. This is believed to be the relic of the 'fireball' phase of the big-bang universe, although other explanations can be invented. The infrared and optical region is unfortunately dominated by atmospheric emission (the airglow), scattered light from interplanetary dust (zodaical light), and light from the stars of the Milky Way.

PROBLEMS

1.1. From the figures given in section 1.1, make a rough estimate of the number of (a) stars (b) galaxies, (c) clusters of galaxies out to the limit of the Mt. Palomar 5m telescope (assume the distances to M31 and the Virgo cluster are typical of the average values).

1.2. Assuming the density of the atmosphere above the ground falls off exponentially with height h:

$$\rho = \rho_0 \exp(-h/h_1),$$

where the scale height h_1 is 8 km, what fraction (by mass) of the atmosphere will still remain above a balloon 40 km up?

2. Our Galaxy, and other galaxies

2.1. INTRODUCTION

In this chapter we look at the structure and evolution of our Galaxy and other galaxies, and introduce the different kinds of 'active' galaxy — radio galaxies, Seyfert galaxies, and quasars.

The main constituents of galaxies are stars and interstellar gas, and the main evolution of a galaxy consist of stars condensing out of gas, undergoing thermonuclear reactions, and finally either cooling down or dispersing much of their material in spectacular explosions.

By tracing out the distribution of populations of stars of different ages in our Galaxy, we can get a good idea of the way that the Galaxy has evolved with time and what the last stages of its formation were like. Other types of galaxy can be explained as similar in structure to our own, and of the same age, but with differing rates of star formation.

In their radio properties, galaxies show an enormous range of power and size. The different types of 'active' galaxy can be understood as representing events of varying degrees of violence in galactic nuclei.

·2.2. STAR FORMATION

The Orion nebula (Fig. 2.1(a)) provides a good example of a gas cloud out of which stars are currently forming. The Trapezium stars have condensed very recently, and one of them is heating up part of the gas to give the visible cloud of ionized hydrogen, or H_{II} region as it is called. Nearby we can see cooler condensations, radiating mainly in the infrared, which are probably even younger 'protostars'.

The formation of a new star starts when some portion of a gas cloud has a slightly higher density than the average for the

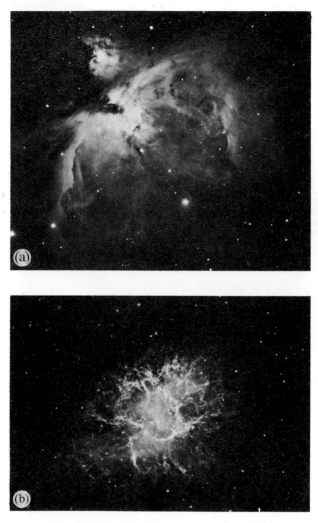

FIG. 2.1. The birth and death of a star: (a) the Orion nebula, part of a gas cloud out of which new stars are forming, (b) the Crab nebula, the relics of a star which exploded as a supernova in 1054 A.D. The core of the star contracted to form a neutron star, which can be seen as a pulsating radio, optical, and X- and γ-ray source. Photographs from Hale Observatories. (Copyright by the California Institute of Technology and the Carnegie Institute, Washington).

cloud. The self-gravitation of this region tends to make it
fall together, and as this happens, what is by now the protostar
heats up. Eventually the central temperature of the condensation
becomes high enough for nuclear reactions to start. For a cloud
composed mainly of hydrogen this happens when the temperature
gets above about 10^7 K: hydrogen then fuses to form helium.
The pressure in the protostar builds up until the pressure
gradient can balance gravity, and the central and surface
temperatures then adjust themselves so that the amount of heat
radiated at the surface is balanced by the amount of energy
generated in the nuclear reactions at the centre. A star is
born.

2.3. THE EVOLUTION OF A STAR

   While a star is 'burning' hydrogen it turns out that the
luminosity L and surface temperature T of the star both depend
only on the mass M of the star (and very slightly on the amounts
of elements other than hydrogen that are present):

$$L \propto M^{3.4}; \; T \propto M^{0.5}. \tag{2.1}$$

Thus stars of different mass lie on a line (the *main sequence*)
in a luminosity-temperature, or *Herzsprung-Russell (HR)*, diagram.
Since colour is a good indicator of temperature, this is usually
used as the horizontal axis by astronomers. Another good
indicator of surface temperature is the type of spectral lines
found in the spectrum of the star, and stars can be classified
according to their *spectral type*. A schematical *HR* diagram
is shown in Fig. 2.2, with the main sequence and the track of a
forming star indicated.

   When most of the hydrogen in the hot central core has been
fused into helium, the star undergoes a dramatic change. The
surface of the star becomes cooler and redder, and the star
grows in size by a large factor, becoming a *red giant*. The core
meanwhile contracts and becomes hotter until helium starts to

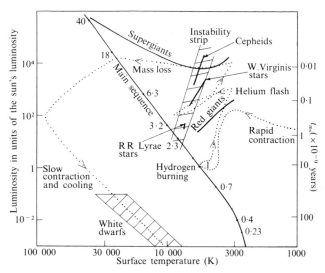

FIG. 2.2. The Herzsprung-Russell diagram, which shows how the luminosity and surface temperature of stars change with time. The dotted curve is the predicted evolution of the sun. The right-hand scale gives the lifetime of stars of different mass on the main sequence (points on the main sequence have been labelled with the appropriate mass in solar units).

fuse into carbon — the *helium flash*. Subsequently thermonuclear reactions progressively fuse carbon into nitrogen, nitrogen into oxygen, and so on, through elements of higher and higher atomic number, up to iron. For elements with atomic number higher than iron the fusion reaction is *endothermic*, i.e. it absorbs energy instead of giving it out, so once the core is composed of iron it can no longer remain in equilibrium. The star either collapses or explodes.

Elements with atomic number higher than iron are formed by a different process — *neutron capture* (see *The properties of nuclei* by G.A. Jones (OPS 12)). The rare-earth elements are formed by the slow irradiation of iron nuclei by neutrons, in red giants, and the radioactive elements are formed by rapid neutron capture in supernovae explosions.

2.4. FINAL STAGES: WHITE DWARFS, NEUTRON STARS, AND BLACK HOLES

The last stages of a star's life depend on its mass, and are not yet completely understood. If the mass is much greater than that of the sun, the star will explode violently as a supernova, showering the interstellar gas with a mixture of all the elements that have been produced in its nuclear reactions. The most spectacular example of a relic of such an explosion is the Crab nebula (Fig. 2.1(b)), the remains of supernova explosion observed by Chinese astronomers in A.D. 1054.

Less spectacular variations and ejections also occur during the later stages of stellar evolution, e.g. the Cepheid and RR Lyrae variable phases (short-period pulsations that occur when the star crosses the 'instability strip' in the HR diagram — see Fig. 2.2), planetary nebulae (caused when a star throws off a spherical shell of gas while it is on the giant branch), and novae (less dramatic outbursts probably arising from mass exchange between the two members of a binary system).

In the very last stage of all, whatever is left of the star after explosions or mass loss must become either (a) a white-dwarf star, in which the density is so high that the electrons are *degenerate* (i.e. crushed together until they are touching), (b) a neutron star, in which the density is even higher and the neutrons are degenerate, or (c) a black hole, where the matter of the star collapses to such a high density that light can no longer escape from it (see section 4.4). Only stars with mass less than about 1.4 times the sun's mass can become white dwarf or neutron stars (in more massive stars the degeneracy pressure is not sufficient to support the star against gravity), so if more massive stars than this fail to eject most of their mass in their eruptive phases, they *must* become black holes.

Before they have cooled off too much, white dwarfs are detectable at optical frequencies, like other stars. Neutron stars, which will tend to be spinning rapidly, can be seen by pulsed radio, optical, and X-ray emission from their rotating magnetospheres (*pulsars*). There is a prominent pulsar in the centre of the Crab nebula, almost certainly the remains of the star that exploded in the 1054 A.D. supernova. Neutron stars can also be seen by X-ray emission from gas heated up by falling at high speed onto the star's (solid!) surface.

The only hope of detecting a black hole is through some indirect effect, e.g. if it is in a binary system with a visible companion. From the period and radius of the orbit of the visible companion we can deduce the mass of the invisible object, and if this is very large deduce that it must be a black hole, especially if X-ray emission from in-falling gas indicates that it is very compact. The X-ray source Cyg X-1 is the best candidate for a binary system containing a black hole, although it has also been suggested that it is a triple system with a neutron star as the compact component.

2.5. THE LIFE HISTORY OF OUR GALAXY

The oldest stars in our Galaxy are $1-2 \times 10^{10}$ years old and are composed of about 73 per cent hydrogen, 27 per cent helium, and almost no other detectable elements. The simplest assumption is that our Galaxy condensed out of a large gas cloud with this composition about $1-2 \times 10^{10}$ years ago. We shall see later that this age and this composition fit in very well with the big-bang picture of the universe (section 3.3 and 5.3).

Structurally our Galaxy can be divided into a disc, a nucleus, and a halo (Fig. 2.3). The stars in the Galaxy can be divided into two main populations: (a) *Population I*, bright young stars found only in the disc, associated with gas, dust, and regions of star formation; (b) *Population II*, faint old halo stars, with very low metal (i.e. elements higher than helium in atomic number) abundance.

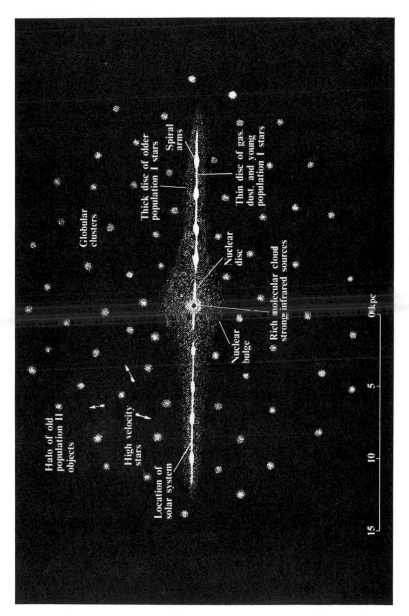

FIG. 2.3. A schematic picture of our Galaxy, seen edge-on.

The different parts of the Galaxy are associated with different phases of its evolution.

The *halo*: *phase* 1

In the halo we find globular clusters containing the oldest stars in the Galaxy, and high-velocity (also old) stars. The age of these clusters can be determined by plotting their *HR* diagram and seeing how much of the main sequence is left. The length of time a star spends on the main sequence is approximately $\propto M/L$, and by eqn (2.1) this means that high-mass stars have much shorter lives than low-mass stars. In a young cluster we will still find quite massive stars on the main sequence, but in an old one only the low-mass ones will still be left there, the more massive ones having already evolved away.

The age of objects in the halo shows that this was the first place where star formation occurred, and the high velocities of the stars suggests that the gas cloud out of which the Galaxy formed was in a state of rapid motion, presumably collapsing. The surface layers of halo stars are found to contain almost no other elements except hydrogen and helium (although some other elements will exist in the interiors of the more evolved halo stars). The time-scale for the collapse of the 'protogalaxy' can be shown to be about 10^8 years.

The thick disc and the nucleus: *phase* 2

The gas that was not used up in phase 1 of the evolution of the Galaxy continued to collapse, and would have formed a concentrated core but for the effect of rotation. The protogalactic cloud must have possessed some angular momentum, since it formed instead a disc, with centrifugal force balancing gravity. This rotating disc was about 2000 light years thick, with a pronounced bulge towards the centre. The second

generation of stars then formed, the more massive ones evolving
rapidly, exploding and spraying the gas with the products of their
nucleosynthesis. Thus in the surface layers of these older
stars of Population I we start to see detectable amounts of
heavy elements. The sun formed during this phase.

The thin disc: phase 3

   The gas still unused, with its admixture of debris from dead
stars, settled down to a thin disc about 300 light years thick.
Stars have continued to form in this disc of gas right up to
the present day, especially in the spiral arms, which probably
represent a spiral wave of star formation rotating slower than
the disc of gas itself. From our position in the disc of the
Galaxy we cannot see the spiral arms, but they can be traced
out through 21-cm line emission from cool, neutral hydrogen (H_I)
We can get an idea of how our Galaxy looks from outside from
photos of external galaxies (Figs. 1.3 and 2.4).

The nuclear disc

   Although we cannot see through to the centre of our Galaxy at
optical wavelengths, due to obscuration by dust, we can do so
in the radio and infrared. In addition to a rotating disc of
gas and stars, we can see a massive cloud of molecular gas,
containing strong infrared sources, probably due to hot dust
grains (10 K < T < 1000 K). There is also evidence of a high-
velocity (several hundred kilometre per second) outflow of gas.
In many ways the nucleus of our Galaxy resembles those of the
more violent objects described in section 2.8, although on a
much weaker scale.

2.6. STRUCTURE AND EVOLUTION OF GALAXIES

   Some examples of external galaxies are shown in Fig. 2.4.

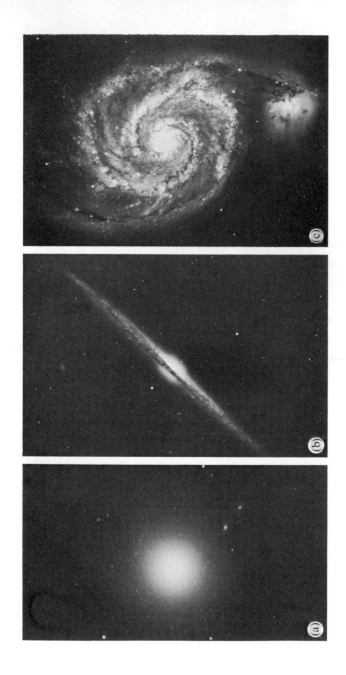

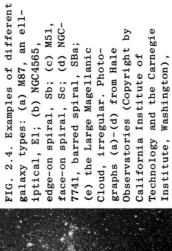

FIG. 2.4. Examples of different galaxy types: (a) M87, an elliptical, E1; (b) NGC4565, edge-on spiral, Sb; (c) M51, face-on spiral, Sc; (d) NGC-7741, barred spiral, SBa; (e) the Large Magellanic Cloud, irregular. Photographs (a)-(d) from Hale Observatories (Copyright by California Institute of Technology and the Carnegie Institute, Washington), (e) by Dr. Whiteoak.

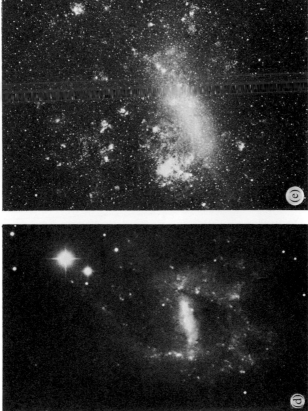

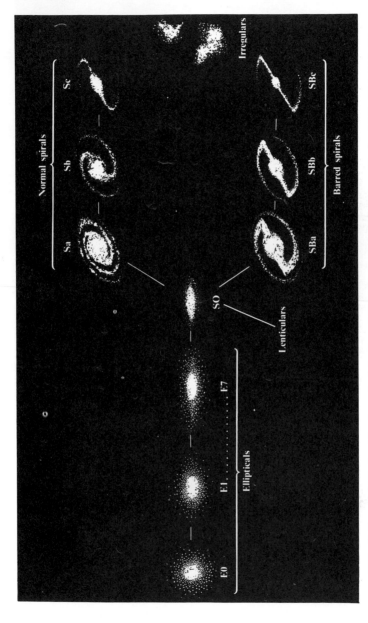

FIG. 2.5. Hubble's 'tuning-fork' classification of galaxies. It is probably not an evolutionary sequence, but one of different rates of star formation.

Hubble introduced a classification of galaxies, the 'tuning-fork' diagram (Fig. 2.5), which is still broadly used, although it is no longer thought to be an evolutionary sequence. The *elliptical* galaxies are classified as E*n*, by their degree of

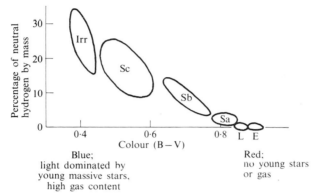

Blue;
light dominated by
young massive stars,
high gas content

Red;
no young stars
or gas

FIG. 2.6. Percentage of galaxy in form of gas, versus the blue-visual (B-V) colour index. This is the difference between the apparent magnitudes (section 3.2) through filters with peak transmission in the blue and visual (i.e. green) parts of the spectrum. From left to right, this could be a sequence of increasing age or, more probably, increasing star-formation rate.

flattening, where $n = 10$ $(a-b)a$ and a and b are the major and minor axes, respectively. They are thought to be spheroidal systems, and this means that we cannot distinguish between a genuinely spherical galaxy and a flattened one seen face-on. However the high proportion of apparently spherical systems shows that we cannot just assume that all ellipticals are flattened systems seen at different orientations.

The S0, or *lenticular,* galaxies seem to be a transitional form between ellipticals and spirals. *Spiral* galaxies are divided into normal spirals and barred spirals, the latter differing only in that their spiral arms start from the end of a prominent bar instead of from the nucleus itself. As we go along the sequence from Sa to Sc for normal spirals, or SBa to SBc for barred spirals, the nucleus becomes less pronounced, the arms become more open and less tightly wound, the proportion of gas increases, and the colour becomes bluer. It is also easier to see individual stars in nearby Sc galaxies, showing that there are more stars with high optical luminosity.

Ellipticals show no sign of gas or recent star formation and are far redder than spirals in colour. Irregular galaxies (Irr), which tend to be very blue and have a high proportion of gas, are placed to the right of the spiral sequence. This trend is summarized in Fig. 2.6, the percentage of mass in the form of neutral hydrogen (as measured by the 21-cm line) against colour.

The blue end of the main sequence corresponds to massive stars, which burn up their fuel very rapidly (10^6 years or so). Hence a blue galaxy must contain young stars, and star formation must still be going on, whereas a red galaxy with no gas (i.e. an elliptical) shows no sign of recent star formation and all its stars could be old. There are examples of all these types of galaxy in the Local Group, as shown in Table 1 (p.4).

The Hubble sequence, however, is almost certainly *not* a sequence of galaxies of different age. Old stars are found in all these galaxy types, including the irregulars. The differing gas content, and proportion of young stars, along the sequence is more likely an indication of *differing rates of star formation*, with all galaxies being of about the same age (about 10^{10} years). Ellipticals formed stars rapidly in phase 1, so no gas was left to form a disc of Population I objects. Irregulars, at the other extreme, have been forming stars at a rather slow rate and still have plenty of gas left.

Clearly the *rotation* of a galaxy must also play a part in determining for example, the degree of flattening of an elliptical, and in the generation of spiral arms and the bars of barred spirals. But the main properties of the Hubble sequence are understood most simply in terms of a 1-parameter (i.e.

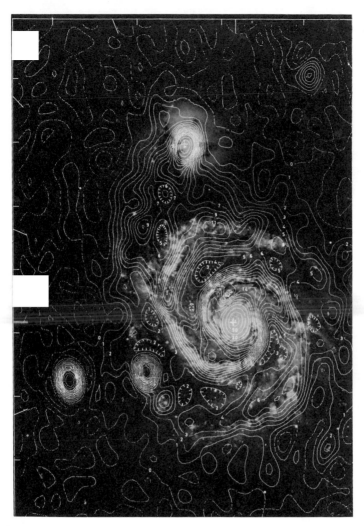

FIG. 2.7. Radio contours superposed on an optical photograph
of the spiral galaxy, M51. Radio map obtained at Westerborg.
(Fig. 1.9(b)).

star-formation rate) family of galaxies of the same age.

An important property for cosmology of elliptical galaxies is that their luminosities and colours are now changing only very slowly, so they can be used as 'standard candles' (see section 6.5).

2.7. RADIO PROPERTIES OF GALAXIES

In addition to starlight we observe from galaxies strong non-thermal synchrotron radiation in the radio band (Figs. 2.7 and 2.8).

Spirals

The radiation from spiral galaxies is concentrated in the disc of the galaxy, with a total power in the range 10^{29}-10^{33}W. There are also weaker sources in the nuclei of spirals, with a typical size of 1000 light years and a power of up to 10^{31}W.

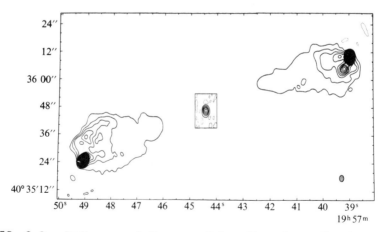

FIG. 2.8. Radio map of the powerful radio-galaxy, Cyg A, showing the characteristic double structure. The weak central component indicates the position of the optical galaxy. Map obtained at Cambridge (Fig. 1.9 (a)). (Hargrave and Ryle (1974), *Mon.Not.R.astr.Soc.* <u>166</u>, 308).

FIG. 2.9. The nearby powerful radio galaxy, Cen A, (NGC 5128) an elliptical with unusual dust lanes across it. Photograph from Hale Observatories. (Copyright by the California Institute of Technology and the Carnegie Institute, Washington).

Ellipticals

The radio emission from elliptical galaxies usually comes from two components symmetrically placed on either side of the galaxy, at total separations of from 10^4-10^7 light years from the galaxy. Total powers range from 10^{33}-10^{39} W and the minimum energies required in magnetic field and relativistic particles range up to 10^{53} J, posing severe theoretical problems. Models have not only to explain this huge energy, which is usually assumed to arise in violent events in the galactic nucleus, but also the strange double structure. Presumably some kind of beaming effect is involved.

A weaker core source is also found in the nucleus of many ellipticals, with typical powers in the range 10^{31}-10^{34} W.

Although the radio properties of galaxies show a wide range of powers and characteristics, it is roughly true that the radio power increases with the optical power (and hence with the mass of the galaxy). Approximately

$$P_{radio} \propto P_{optical}^2. \tag{2.2}$$

Fig. 2.9 shows the galaxy NGC 5128, the most powerful radio emitter in our neighbourhood.

2.8. 'ACTIVE' GALAXIES AND QUASARS

Three types of object show evidence of violent, transient activity. In order of increasing violence of the activity, these are Seyfert galaxies, N-galaxies, and quasars. It is hard to draw a sharp line between the different classes, since each type merges into the next. In each case we are seeing violent activity in the nuclei of galaxies, although in the case of quasars it is not at all certain that we are looking at stellar systems.

Seyfert galaxies

These are galaxies (usually spirals) with an intense, often variable, star-like nuclear region. The spectra of these nuclei show very strong, broad emission lines, arising from hot gas rather than from stars. Colours are bluer than average, suggesting the presence of a non-thermal component to the optical continuum. In some cases there is evidence for the violent ejection of material, at velocities of several thousand kilometres per second.

Seyfert nuclei have been found to be unusually strong infrared

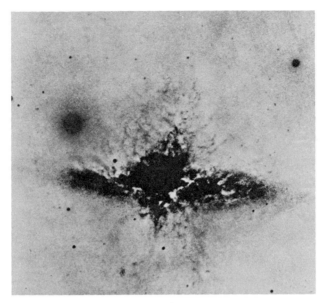

FIG. 2.10. The irregular galaxy, M82, showing filaments believed to originate in an explosion in the nucleus. The galaxy is a very strong emitter of far-infrared radiation and a moderately strong radio and X-ray emitter.

sources, but they are only slightly more active than normal galaxies at radio wavelengths.

Fig. 2.10 shows the irregular galaxy M82, which resembles a Seyfert in being a strong infrared emitter, and which shows evidence of a violent explosion in its nucleus.

N-galaxies

These are elliptical galaxies which are unusually luminous radio-wise and which have a strong, compact optical nucleus. In some cases the nucleus is so strong that the rest of the galaxy can only just be seen. Their colours are distinctly bluer than the typical elliptical, probably due to non-thermal optical continuum radiation. Some emission lines are broad, but their spectra are less excited than those of Seyferts.

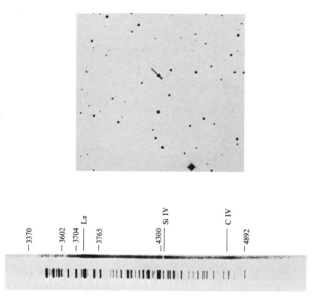

FIG. 2.11. Finding-chart and spectrum of the quasar, 1331+170.
The arrow points to the quasar. The two broad emission lines
in the spectrum are the Lyman-α line of hydrogen and a prominent
carbon line, and there are several absorption lines. (Strittmatter
et al (1973), *Astrophys. J.* **183**, 767).

Quasars

These include radio sources associated with quasi-stellar
optical objects (QSRs) and also quasi-stellar optical objects
with no evidence for radio emission (QSOs). Their spectra show
broad emission lines from hot gas only (there are no stellar
lines) (see Fig. 2.11). For a few quasars, distances can be
determined by association with neighbouring clusters of galaxies.
The optical powers range from a few times up to a thousand times
that of a normal luminous elliptical galaxy, assuming their
red-shifts are cosmological (see section 3.3). Their colours
are usually very blue.

The radio sources associated with quasars fall into two

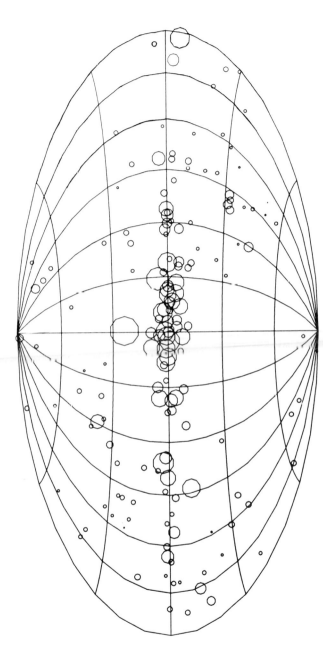

FIG. 2.12. The distribution of the X-ray sources in the third Uhuru catalogue (see Table 6),in galactic coordinates. The size of the circles is proportional to the X-ray flux. The strongest sources, associated with compact objects (white dwarfs, neutron stars, and black holes) in binary systems in our Galaxy, are concentrated towards the galactic plane. Most of the identified sources at high galactic latitude are associated with rich clusters of galaxies, a few with active galaxies and quasars.

types: (a) double sources similar to those of elliptical
galaxies; (b) compact sources, variable on a time-scale of a
month or less. The latter class of quasar also sometimes show
violent optical variations. It is remarkable that objects not
much larger than the solar system, putting out up to a thousand
times as much energy as the whole Milky Way, can vary in their
power so rapidly.

The simplest picture for quasars assumes that they represent
galaxies in which the source in the nucleus has become so
luminous that the outlying galaxy can no longer be seen. Some
are slightly fuzzy optically, perhaps due to a surrounding
galaxy.

2.9. CLUSTERS OF GALAXIES

While most galaxies may well fall in groups similar to the
Local Group, a small proportion are to be found in rich and
spectacular clusters of galaxies like the Virgo cluster,
containing many thousands of members and with total masses
ranging up to 10^{15} $M_\odot$ ($M_\odot$ = Mass of the sun). Although it is
not yet well established, there is probably also a substantial
amount of intergalactic gas within these rich clusters. Two
lines of evidence which point to this are 'tail' radio galaxies,
in which radio-emitting plasma appears to have been swept
backwards by the motion of the galaxy through gas, and strong
X-ray emission from rich clusters probably due to hot gas (see
Fig. 2.12).

PROBLEMS

2.1. Describe the appearance of a galaxy like our own
      (a) 10^8 years, (b) 10^9 years, (c) 10^{11} years after
      it started to form.

2.2. Given that an object looks stellar if it subtends an
      angle smaller than $1''$ of arc, at what distance would
      our Galaxy look stellar?

3. The empirical basis for cosmological theories

3.1. INTRODUCTION

We can make progress towards answering the questions 'what is the structure of the universe?' and 'how has the universe changed with time?' only if we believe that the universe has a simple overall structure.

In our immediate locality the universe has an exceedingly complex structure. If we had to construct a mathematical model of the human brain, and had as little observational information about it as we have about the universe, we would hardly know where to start. However, a human brain is a highly untypical region of the universe. For one thing its density is about 10^{30} times the present average density of matter in the universe.

But suppose we start looking at things on the large scale — on the scale of the distances we have been mentioning in Chapters 1 and 2. Then the earth, and all its structure and detail that is so important to us, becomes a minute speck of dust. On this scale the universe may start to look simple. Before we can start to discuss what sort of properties the universe might have when looked at in this way, we must establish a cosmological distance scale.

3.2. THE DISTANCE SCALE

Cosmological distance can be measured in a variety of ways, all involving the observation of light signals. Unfortunately the most reliable methods can be used only over a short range of distances. As we look further out, we are forced to use progressively less reliable methods.

Radar techniques

These can be used to measure the distances of only our nearest neighbours in the solar system.

Parallax

The apparent change in direction of a source as the earth goes round the sun gives distances for the nearest few hundred stars. The distance is measured in *parsecs*; a distance is 1 parsec (pc) (= 3.09×10^{16} m = 3.26 light years) if the mean radius of the earth's orbit subtends $1''$ of arc. Distances measured in this way give us the luminosities of nearby stars of different type. The apparent brightness of certain types of star can then be used as a distance indicator.

Luminosity distance

The luminosity distance d_{lum} is defined as the distance an object would appear to have if the inverse-square law held exactly (cf. eqn (1.4)):

$$d_{lum} = (P/S)^{1/2}, \tag{3.1}$$

where P is luminosity and S is the flux density of radiation. The luminosity distance provides a ladder of distance using objects ranging from moderately luminous variable stars to whole clusters of galaxies. All that is needed is a class of source (a *standard candle*) with not too great a spread in luminosity P. At visible frequencies astronomers have got into the rather odd habit of measuring flux density and luminosities in terms of *apparent magnitude m* and *absolute magnitude M*, where the magnitude scale is such that there are 5 magnitude steps per range of brightness of 100, with the additional peculiarity that the greater the magnitude the fainter the source. Thus

$$m = A - 2.5 \lg S. \tag{3.2}$$

This scale was adopted in the nineteenth century to agree approximately with the brightness classification given by Hipparchus for bright stars. The constant A depends on the range of the spectrum being used. The absolute magnitude M is defined as the magnitude the source would have at a distance of 10 pc, so

$$M = A - 2.5 \lg P + 87.45. \tag{3.3}$$

The absolute magnitudes of the Local Group of galaxies were shown in Table 1 (p.4).

The *distance modulus* is then defined as $m - M$, and it can easily be seen from eqns (3.1) - (3.3) that it is related to luminosity distance by

$$m - M = 5 \lg d_{lum} - 87.45, \tag{3.4}$$

where d_{lum} is in metres.

Table 3 shows how luminosity distances for different types of object give a ladder of distance reaching out to cosmological dimensions. However, it is a ladder that becomes shakier the further out it reaches. Only for galaxies in the Local Group (see Table 1, p.4) can distances be measured by many independent methods

Diameter distance

The diameter distance d_{diam} is defined by assuming the apparent angle θ subtended by an object of intrinsic size l varies inversely with distance:

$$d_{diam} = l/\theta. \tag{3.5}$$

It can be obtained for only a few nearby stars, but can be used for more extended objects, ranging from planetary nebulae and globular clusters to clusters of galaxies. What is needed here is a class of object with a small dispersion in the intrinsic size l. Table 3 shows the contribution of diameter distance methods to the cosmological distance scale.

TABLE 3
The distance scale

Distance indicator	Luminosity distance	Diameter distance	Range
Variable stars	RR Lyrae		150 kpc
	Cepheids		4 Mpc
Bright stars	Brightest stars in galaxies		10 Mpc
Eruptive stars	Novae		8 Mpc
	Supernovae		400 Mpc
Extended objects in galaxies		Planetary nebulae	500 kpc
	Globular clusters	Globular clusters	10 Mpc
	H_{II} regions	H_{II} regions	25 Mpc
Galaxies	Luminosity class of spiral galaxies		100 Mpc
	Brightest galaxies in clusters	Largest galaxies in clusters	4000 Mpc
Clusters of galaxies	Luminosity function of cluster galaxies	Cluster core size	2000 Mpc

Some of the uncertainties which arise from these distance methods are well illustrated from our normal visual experience. The human eye uses parallax distance out to about 5 m, and diameter distances beyond that. However, on the road at night the only way of deciding the distance (and hence the speed) of a distant motor bike is by luminosity distance, but there is the possibility of confusion with a (much nearer) bicycle. For a car at night diameter distance can be estimated from the angle subtended by the head or tail lights, but there is considerable error involved due to the great range in width between a bubble-car and a juggernaut.

For objects travelling much slower than the speed of light in a Euclidean space all these definitions of distance give the same answer. In a curved space, dicrepancies between the different methods may help to determine the curvature.

3.3. THE RED-SHIFT

If we look for some particular spectral lines in the sun's spectrum, e.g. the sodium doublet at 5900 Å, we will find that the wavelengths of these and all the other spectral lines are in general shifted by a small amount; this is due to various factors.

(a) The rotation of the earth about its axis results in a small Doppler shift of spectral lines, depending on the time of day. In the morning the observer has a component of motion towards the sun, so the frequencies appear to be shifted to higher frequencies, towards the blue end of the spectrum. Similarly in the afternoon there is a shift towards the red. The amount is minute, at most 1.4 parts in a million.

(b) The sun's own rotation, and circulatory motions in the sun's surface layers, cause a shift of the same order of magnitude, the direction depending on which part of the sun you are looking at.

(c) According to Einstein's general theory of relativity there will be a *gravitational red-shift* by an amount $2GM/Rc^2$, where M and R are the mass and radius of the sun, and G is the gravitational constant. Essentially the photons have to do work in climbing out of the sun's gravitational field, lose energy, and so end up at a lower frequency than when they set off. The fractional shift

$$\Delta \nu / \nu_e = (\nu_e - \nu_o)/\nu_e,$$

where ν_e and ν_o are the emitted and observed frequencies, again turns out to be about one part in a million.

(d) Now imagine that we look at another star in our own Galaxy. In addition to the shifts (1) - (3) above, there may be a far more significant shift due to the motion of the star round the Galaxy. The sun, like the majority of stars in the plane of the Milky Way, is moving in a roughly circular orbit round the Galaxy, and has a circular velocity of 250 km s^{-1}. Some stars are weaving in and out of the galactic plane with speeds of the same order. The net result is that frequency shifts up to

$$\left| \Delta\nu/\nu_e \right| \sim 10^{-3} \qquad (3.6)$$

may be expected.

Finally, what happens when we look at the integrated spectrum of another galaxy? First, since the light from a galaxy is made up of light from many stars, we will expect the spectral lines to be spread out by about the amount given by eqn (3.6). Secondly, since the galaxies have some relative motion with respect to each other (nearby ones affect each other gravitationally, for example), we would expect to find some moving towards us, some moving away, i.e. some blue-shifted, some red-shifted.

We do indeed find this for the nearest galaxies, e.g. the members of the Local Group of galaxies mentioned in section 1.1. Typical (so-called 'peculiar') velocities are a few hundred kilometres per second. But when we start to look at more distant galaxies, as determined by the methods of section 3.2, a remarkable fact emerges (discovered by Hubble and Lundmark in the 1920s and strongly confirmed by the more recent work of Sandage and coworkers). Almost all the frequency shifts are red-shifts and *the red-shift increases linearly with distance.* The slope of this line determines a constant with the dimensions of distance, which we shall call the *Hubble distance* and write as $c\tau_o$, where τ_o is the *Hubble time* (the time for light to travel a Hubble distance). If we write z for the red-shift, i.e.

$$z = \Delta\nu/\nu_e = (\nu_e - \nu_o)/\nu_e,$$

then

$$z \simeq d/c\tau_o + v_{pec}/c, \qquad (3.7)$$

where d is the distance of the galaxy, v_{pec} is its peculiar velocity in the line of sight, and d/c is the *cosmological red-shift*.

Except for nearby galaxies, the peculiar velocity can be neglected, so strong is the effect of the cosmological red-shift. *If* (and this is a big 'if') the red-shift is a Doppler shift, then the galaxies are receding from us in every direction, with a velocity that increases with distance from us (Figs. 3.1 and 3.2). In fact in this case

$$v \simeq zc \simeq d/\tau_o, \text{ provided } v \ll c,$$
$$= H_o \, d, \tag{3.8}$$

where $H_o = \tau_o^{-1}$ is called the *Hubble constant*. If τ_o is given

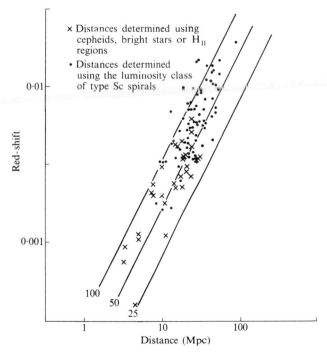

FIG. 3.1. The red-shift-distance relation, showing that the red-shift of galaxies is proportional to distance. The solid lines correspond to different values of the Hubble constant in $km \, s^{-1} \, Mpc^{-1}$.

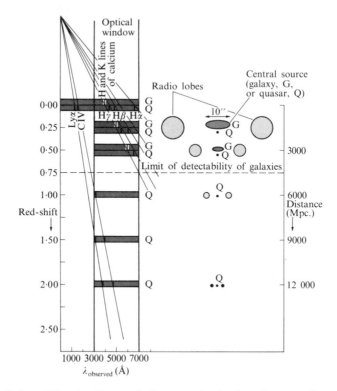

FIG. 3.2. Illustration of the way typical galaxy and quasar spectral lines are red-shifted across the visible wavelength band, and of the way this red-shift increases with distance, indicated by the angular size and separation of the radio components and by the angular size of the galaxy (the quasar appears point-like at all distances).

in light years, H_o is in (light years)$^{-1}$. However, you will often find it given in units of km s^{-1}/Mpc^{-1}. The current popular value is H_o = 50 km s^{-1}/Mpc^{-1}, corresponding to $\tau_o \approx 2 \times 10^{10}$ years and $c\tau_o \approx 2 \times 10^{26}$ m $\approx$ 6000 Mpc.

Clearly the time for a galaxy at distance d travelling at velocity v to travel a further distance d is, by eqn (3.8), $d/v = \tau_o$, independent of d. Thus the Hubble time is a measure of the expansion time of the universe, the time for the universe to double its size expanding at the present rate.

Of course, if the red-shift is *not* a Doppler shift, then the Hubble time has no such significance.

3.4. ISOTROPY
Galaxies

We saw in Chapter 1 that the distribution of the nearest galaxies on the sky is far from isotropic, due to our membership of the Local Group of galaxies and proximity to the Virgo cluster. But on the large scale the distribution of galaxies seems to be fairly isotropic, once allowance has been made for the obscuration due to dust in our Galaxy. The exact nature of these tiny (0.1 μm diameter) dust grains responsible for absorbing and scattering light within the Galaxy is uncertain, but they probably consist of graphite cores surrounded by mantles of 'dirty' ice (i.e. contaminated with metals) or silicates. The grains appear to be distributed throughout the gas pervading the thin disc of our Galactic plane. The result is that a line of sight close to the plane passes through a much longer column of dust than a direction normal to the plane. For a uniform distribution of dust throughout a slab of thickness $2h$, the path length that light from outside the slab has to traverse is $h \csc b$, where b is the galactic latitude of the

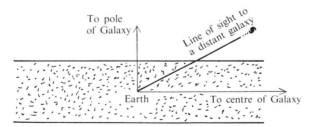

FIG. 3.3. Part of the galactic disc near the earth. The line of sight to a distant galaxy has to traverse a distance $h \csc b$ of the dusty gas in the disc, where $2h$ is the thickness of the disc and b is the galactic latitude.

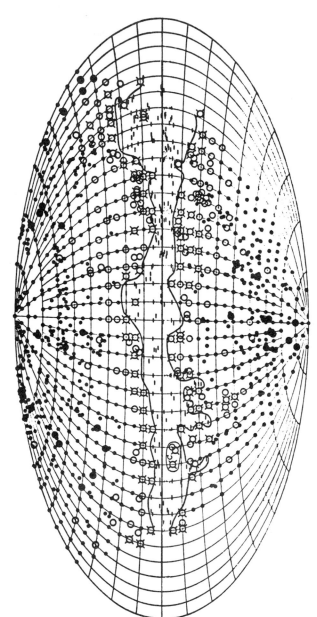

FIG. 3.4. The distribution of Hubble's galaxy counts to 20th magnitude on the sky, in galactic coordinates. Large filled circles denote fields where an above-average number of galaxies was counted, large open circles where the number was below average, and a dash where none at all could be seen. Near the galactic plane there is a 'zone of avoidance' where almost no galaxies are seen, owing to obscuration by dust (cf. Fig. 1.4).

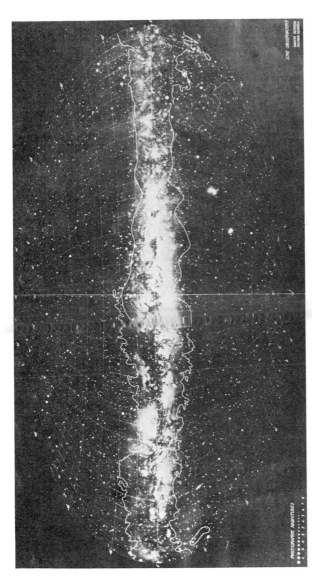

FIG. 3.5. Composite photograph of the Milky Way by the Lund observatory, with a contour of constant number of galaxies per square degree from the Lick and Harvard counts superposed. The contour corresponds to a low number per square degree to illustrate how the zone of avoidance follows the regions of high star (and hence gas and dust) density in the Galaxy.

source (Fig. 3.3). In fact the dust concentration varies with distance from the Galactic plane in a roughly inverse exponential way. However, in the approximation that the concentration depends only on distance from the plane over the line of sight, the cosec law still applies.

The flux from a distant source is then reduced from S_o to

$$S = S_o \exp (-kh \cosec b), \qquad (3.9)$$

where k denotes the extinction per unit length (i.e. $dS/dx = - kS$). In terms of magnitudes,

$$m = m_o + 1.086 \ kh \cosec b. \qquad (3.10)$$

Thus if we have a telescope capable of seeing galaxies down to a particular value of m, the corresponding limiting value of m_o will depend on b. For smaller values of b we cannot see such distant objects as we can towards the galactic pole ($b = 90^{\circ}$). This in turn means that we will not see so many galaxies per square degree.

The actual distribution of galaxies, in galactic coordinates (analogous to latitude and longitude, with the Galactic plane as equator), is shown in Fig. 3.4. This is derived from Hubble's counts of the galaxies per square degree in different directions. Within the 'zone of avoidance' close to the galactic plane, no galaxies are seen at all (see also Fig. 3.5).

Naturally it is hard to say much about the isotropy of the universe from these data, but once a correction is applied for obscuration the distribution is certainly *compatible with isotropy*, apart from the tendency already mentioned for galaxies to occur in clusters.

Clusters of galaxies

Abell has made a catalogue of 'rich' clusters of galaxies, i.e. those with a specified minimum number of bright members.

Their distribution on the sky is again compatible with isotropy, once the correction for obscuration has been applied.

Radio sources

The vast majority of the bright radio sources in directions away from the galactic plane are extragalactic, either galaxies or quasars. There is no galactic extinction of radio waves, except at very low frequencies, so we can test directly for isotropy. So far, down to the faintest sources at present detectable, the distribution of radio sources on the sky is isotropic to 10 per cent. Such discrepancies as have arisen between one survey and another have usually turned out to be due to errors in the flux-density scale. Some controversy remains about possible anisotropies in the distribution on the sky of quasars.

The microwave background radiation

So far we have looked at the distribution in the sky of *matter* and although no clear anisotropy has emerged, the degree of isotropy we can claim is no better than 10 per cent. Far more significant evidence for isotropy of the universe on the large scale comes from the microwave background radiation (section 1.8). At a wavelength of 3 cm, this radiation has been shown to be isotropic to 0.1 per cent.

Whether this radiation is the relic of the 'fireball' phase of a big-bang universe (see p.84) or is due to discrete sources of microwave radiation in galaxies, this isotropy dates from long times in the past, and places severe limits on any anisotropic models of the universe.

3.5. UNIFORMITY

Newton realized that if the stars were distributed with uniform number density throughout the universe, as had been proposed by

Digges and Bruno, then this could be tested by counting the number of stars as a function of their observed flux.

For if a set of sources all of the same luminosity P are distributed uniformly with number density η, then the number per steradian out to distance r is

$$N(r) = \eta r^3/3,$$

while their observed flux

$$S = P/r^2.$$

The number of sources per steradian brighter than S is therefore

$$N(S) = (\eta P^{3/2}/3)S^{-3/2} \tag{3.10}$$

or

$$\lg N = A - 1.5 \log S \tag{3.11}$$

$$= B + 0.6 \, m,$$

by eqn (3.2), where A and B are constants.

Newton's attempts to apply this test of uniformity failed

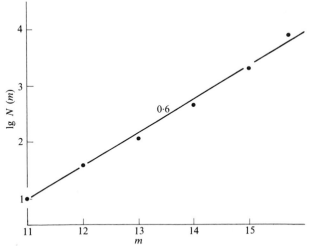

FIG. 3.6. The number of galaxies per steradian as a function of photographic magnitude, showing, at least for the relatively bright magnitudes included here, good agreement with the expected d lg $N/$dm = 0.6 relation. Fields near the Virgo cluster and Local Supercluster plane (see Epilogue p.139) have been omitted.

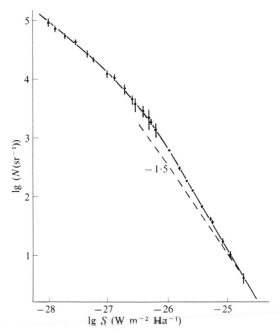

FIG. 3.7. Radio-source counts, lg N against lg S, at a frequency of 408 MHz. Even for the brightest sources, the curve rises more steeply than the -1.5 predicted (eqn (3.11)) if all red-shift and evolutionary effects are neglected.

because he had no adaquate way of estimating the fluxes of stars. Later, William Herschel used this test to show that the stars of our Galaxy are in fact distributed in a disc.

For distant galaxies we do not expect eqn (3.11) to hold exactly, since the red-shift, whatever its interpretation, will affect the observed fluxes. But for brighter galaxies, eqn (3.11) does indeed hold approximately (Fig. 3.6). For radio sources, on the other hand, a significantly steeper slope is found (Fig. 3.7), suggesting that the number density or luminosity of these sources was greater in the past. However, red-shift effects cannot be neglected in interpreting radio-source counts.

3.6. OLBERS' PARADOX

Halley, and later Cheseaux and Olbers, realized that important cosmological information is contained in the fact that the sky is dark at night.

Suppose a population of sources of luminosity P has a number density η. Consider a spherical shell of radius r, thickness dr, centred on the earth. The number of sources in one steradian of the shell is

$$dN = \eta r^2 dr,$$

and their flux

$$S = P/r^2.$$

The intensity of the integrated light from these sources is

$$dI = SdN = \eta Pdr$$

and if we add the contribution from shells with $0 \leqslant r \leqslant R$, the total intensity

$$I = \eta \, PR. \tag{3.12}$$

Clearly if we let $R \rightarrow \infty$, then $I \rightarrow \infty$. (Actually the stars would eventually start to block out more distant light, so the intensity would tend to the average surface brightness of the stars - something comparable to the surface brightness of the solar disc.) Since the intensity of the integrated light from galaxies has to be much less than the intensity of the Milky Way, we can deduce that we receive light only from galaxies with a distance R less than some maximum distance R_{max}. The simplest reason for this would be that the universe is of finite age. An infinitely old expanding universe would be another way of resolving this paradox.

This 'paradox' can also be stated in a thermodynamic form: why is the universe so cold? Again, in a universe of finite age, we see that there may not have been time for the stars to heat the rest of the matter up.

3.7. EVIDENCE FOR A UNIVERSE OF FINITE AGE

Several pieces of evidence suggest that we may live in a universe of finite age.

1. The proportions of different isotopes of radioactive elements allow quite accurate dating of different materials. For example, the oldest rocks on the earth, the moon, and in meteorites have ages of about $(4.3 \pm 0.3) \times 10^9$ years, in good agreement with the age of the sun estimated from calculations of its structure and evolution, 5×10^9 years. Applied to the material in our Galaxy as a whole, radioactive dating gives an age of 1.1-1.8×10^{10} years, again agreeing well with estimates of the ages of the oldest stars, from stellar evolution theory 1-2×10^{10} years.

2. The ages of other nearby galaxies are of the same order, independent of galaxy type.

3. If the cosmological red-shift is due to expansion of the universe, then the age of the universe will be of order the Hubble time, 2×10^{10} years. The similarity in the ages of galaxies in our neighbourhood and the expansion time-scale of the universe strongly suggests a universe of finite age in which galaxies formed early on. However it is hard to eliminate the possibility that the part of the universe we have studied so far is merely some local fluctuation, or that the universe has had a cyclical history.

3.8. EVIDENCE FOR A 'FIREBALL' PHASE

The existence of an early phase of the universe's history in which radiation played a dominant role (the 'fireball' phase) is the simplest explanation of the 2.7 K black-body background radiation.

A second piece of evidence for such a phase is that the stars in our Galaxy appear to have been formed with an initial composition of 73 per cent hydrogen, 27 per cent helium, by mass. Almost all the remaining elements can be formed within the interiors of stars, so it is natural to suppose that the universe started as pure hydrogen and that the helium was formed in some pre-stellar phase, i.e. the fireball. While it is possible to construct a theory of the early stages of our Galaxy's history in which very massive stars formed, evolved rapidly, and exploded, dispersing helium, the most natural explanation is that the helium was formed through nuclear reactions in the early fireball phase of the universe.

PROBLEMS

3.1.   Use eqn (3.10) to work out how many magnitudes an object in the direction of the galactic pole is dimmed, given that there are about 20 magnitudes of extinction to the galactic centre (10 kpc distant) and that $h = 150$ pc.

3.2.   What is the limit to which diameter distance can be used by (a) the human eye (lens diameter $D = 5$mm) working at $\lambda = 5000$ Å, (b) a 100-m diameter radio telescope working at $\lambda = 20$ cm (Assume that the limit of resolution is $2\lambda/D$ rad.)

4. The big-bang models

4.1. THE SUBSTRATUM AND FUNDAMENTAL OBSERVERS

To describe the properties of a gas we do not need to study the
behaviour of individual atoms and molecules. Instead we define
various macroscopic quantities — density, pressure, temperature -
and study the relations between these.

In the same way, we make no attempt to incorporate individual
galaxies, or clusters of galaxies, into our description of the
universe as a whole. We imagine the matter in the universe
smeared out into an idealized, smooth fluid, which we call the
substratum.

We call an observer who is at rest with respect to this
substratum *a fundamental observer*. If the substratum is in
motion, then we say that fundamental observers are *co-moving* with
it. We are interested in obtaining the picture that fundamental
observers have of the universe as a function of time. We call
such a set of pictures a *cosmological model*.

In practice we will often identify fundamental observers with
galaxies, and will treat an observer at the centre of our Galaxy
as one, although in fact individual galaxies will have some
'peculiar' velocity with respect to the substratum. (Indeed we
may now be close to being able to measure the peculiar velocity
of our Galaxy through observations of the microwave background
radiation.) But for the moment we shall assume that when we
have corrected for the rotation of the earth, its orbit round
the sun, and the motion of the whole solar system round the
Galaxy, then we are receiving a fundamental observer's view of
the universe.

4.2. THE COSMOLOGICAL PRINCIPLE

It is evident that in the post-Copernican era of human history, no well-informed and rational person can imagine that the earth occupies some special position in the universe. We shall call this profound philosophical discovery the *Copernican principle*, although the first clear statement of it is due to Giordano Bruno. The discovery of millions of stars like the sun, of other possible planetary systems, and of galaxies similar to our own Galaxy, all help to convince us of the truth of the Copernican principle. Bruno himself knew of no such evidence, so his affirmation was more of a poetic, psychological, and even political truth.

We now make a much more powerful assumption, the *cosmological principle*: the universe as seen by fundamental observers is homogeneous and isotropic.

By *homogeneity*, we mean that every fundamental observer sees the same general picture of the universe as a function of time. Every fundamental observer is equivalent to every other and, in particular, the universe as seen by any fundamental observer looks the same as the universe as seen from earth. The hypothesis of homogeneity can never be strictly tested, for even if advanced civilizations in distant galaxies transmitted their cosmological knowledge to us, it would always be out of date by the time it arrived. The power of the hypothesis is that our own observations are all that we need to test a cosmological model.

By *isotropy*, we mean that the universe looks the same to a fundamental observer whichever direction in the sky he looks. Observations in one direction only are sufficient to test a cosmological model. The hypothesis is only roughly (~10 per cent) tested for the matter in the universe, but it is very accurately (<0.1 per cent) satisfied for the microwave background radiation (section 3.4).

The cosmological principle satisfies the Copernican principle in almost the strongest possible way. Considerable theoretical effort has gone into investigating models which are homogeneous but anisotropic. It can be shown that isotropy, together with the Copernican principle, implies homogeneity.

An immediate consequence of homogeneity is the existence of a universal *cosmical time*, which we shall denote by t. For since all observers see the same sequence of events in the universe, they can synchronize their clocks by means of these events.

4.3. NEWTONIAN COSMOLOGY

Newtonian dynamics and gravitation can be used to construct models of the universe satisfying the cosmological principle. The cosmical time t can then be identified with the uniform ever-flowing universal Newtonian time.

Consider a fundamental observer O. He sets up coordinates with himself at the origin, and observes the physical properties of the matter at a general point P at time t, the position of P being given by $\overrightarrow{OP} = \underline{r}$. Naturally O finds that the velocity, pressure, and density at P are functions of position and time, $\underline{v} \; (\; \underline{r}, \; \underline{t})$, $\rho \; (\underline{r}, \; \underline{t})$, $p \; (\underline{r}, \; \underline{t})$.

Now consider a second fundamental observer O', distant $\overrightarrow{O'P} = \underline{r}'$ from P. He finds $\underline{v}' \; (\underline{r}', \; \underline{t})$, $\rho \; (\underline{r}', \; \underline{t})$, $p \; (\underline{r}', \; \underline{t})$. Note that in Newtonian theory, the density and pressure at any point are the same for all observers (we say they are *invariants* with respect to change of coordinate system). To satisfy the cosmological principle we now demand that $\underline{v}'$, ρ, and p should be the same functions of $(\underline{r}', \; t)$ as $\underline{v}$, ρ, and p are of $(\underline{r}, \; t)$, otherwise O' and O would have different pictures of the universe.

At a fixed time t, let $\overrightarrow{OO'} = \underline{a}$, so that the velocity of O' as

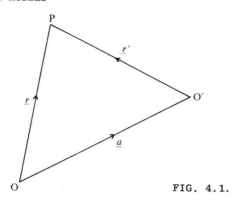

FIG. 4.1.

seen by O is $\underline{v}$ ($\underline{a}$) (we shall ignore the dependence on t for the moment). From Fig. 4.1 we can see that

$$\underline{r}' = \underline{r} - \underline{a},\qquad(4.1)$$

by the vector law of addition applied to the triangle OO'P(Fig. 4.1). It also follows that

$$\underline{v}'(\underline{r}') = \underline{v}\ (\underline{r}) - \underline{v}\ (\underline{a}),\qquad(4.2)$$

by the vector combination of velocities.

Now eqn (4.1) implies that

$$\underline{v}'\ (\underline{r}') = \underline{v}'\ (\underline{r} - \underline{a}),\qquad(4.3)$$

and homogeneity requires that

$$\underline{v}'\ (\underline{r} - \underline{a}) = \underline{v}\ (\underline{r} - \underline{a}),\qquad(4.4)$$

since O' and O have to see exactly the same picture of events. Collecting eqns (4.2) - (4.4) together:

$$\underline{v}\ (\underline{r} - \underline{a}) = \underline{v}\ (\underline{r}) - \underline{v}\ (\underline{a}).\qquad(4.5)$$

Homogeneity also tells us that the way density and pressure vary with position as measured by O must be the same as they vary with position as seen by O', i.e.

$$\rho\ (\underline{r}') = \rho\ (\underline{r});\ p\ (\underline{r}') = p\ (\underline{r}),$$

and then eqn (4.1) implies that

$$\rho\ (\underline{r} - \underline{a}) = \rho\ (\underline{r});\ p\ (\underline{r} - \underline{a}) = p\ (\underline{r}).\qquad(4.6)$$

Since O, O', P are arbitrary so also are $\underline{r}$, $\underline{r}'$, $\underline{a}$, and therefore eqns (4.6) imply that p and ρ must be independent of position.

The general solution of the three equations (4.5) can be shown to be

$$v_i (\underline{r},t) = \sum_{k=1}^{3} a_{ik} (t) x_k, \text{ for } i = 1,2,3, \qquad (4.6)$$

where v_i denotes the ith component of $\underline{v}$; x_k denotes the kth component of $\underline{r}$; and $a_{ik} (t)$, $i,k = 1,2,3$, are nine arbitrary functions of the time t. (Writing out the first of the three equations (4.6) in full:

$$v_1 (x_1,x_2,x_3,t) = a_{11} (t)x_1 + a_{12} (t)x_2 + a_{13} (t)x_3.)$$

For the flow field represented by eqn (4.6) to be isotropic we must have

$$a_{ik} = 0, i \neq k,$$

and

$$a_{11} = a_{22} = a_{33}$$
$$= H(t),$$

say. Therefore

$$v_1 = H(t)x_1, \quad v_2 = H(t)x_2, \quad v_3 = H(t)x_3$$

or

$$\underline{v} = H(t) \underline{r}. \qquad (4.7)$$

This equation tells us that the velocity of any particle moving with the substratum is either zero or is directed radially away from or towards us, with a velocity proportional to distance. In other words we have a natural explanation of the Hubble law, with the cosmological red-shift being interpreted as a Doppler shift. And since this velocity field satisfies the cosmological principle, any other fundamental observer sees exactly the same picture: every particle moving with the sub-stratum has a purely radial velocity, proportional to its distance from him.

Eqn (4.7) can be integrated, since $\underline{v} = d\underline{r}/dt$, by writing

$$H(t) = \frac{1}{R(t)} \frac{dR(t)}{dt},$$

giving

$$\frac{d\underline{r}}{dt} = \frac{1}{R(t)} \frac{dR}{dt} \underline{r},$$

which has the solution

$$\underline{r} = R(t) \times \text{a constant vector}$$

$$= \frac{R(t)}{R_0} \underline{r}_0 \qquad (4.8)$$

where $R_0 = R(t_0)$, so that $\underline{r} = \underline{r}_0$ at epoch $t = t_0$. $R(t)$ is called the *scale factor* of the universe, since as time proceeds all distances are simply scaled up by this factor (Fig. 4.2). The

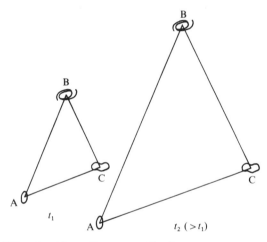

FIG. 4.2. Illustrating the way all distances are scaled up by the same factor as time changes.

only motions permitted by the cosmological principle are a simple isotropic expansion or contraction of the whole universe.

Since the volume will be proportional to $R^3(t)$, the density of matter

$$\rho(t) \propto R(t)^{-3}, \qquad (4.9)$$

i.e.

$$\rho(t) = \rho_0 R_0^3 / R^3(t), \qquad (4.10)$$

where $\rho_0 = \rho(t_0)$. We shall generally take the reference epoch

$t = t_0$ to be the present epoch, and when quantities have a subscript zero it will denote their values at the present epoch. Since at the present epoch the energy density of radiation in the universe has a negligible dynamical effect, we shall for the moment take the pressure p to be zero. This assumption will break down in the early stages of the big bang (see Chapter 5).

To find the form of the function $R(t)$, consider a spherical shell of particles co-moving with the substratum, centred on the observer O. From eqns (4.7) and (4.8), particles A, B, C, D on this spherical shell are receding with velocity $\dot{R}\underline{r}_0/R_0$, and have acceleration $\ddot{R}\underline{r}_0/R_0$, where $\dot{R}$ and $\ddot{R}$ denote $(dR/dt)(t)$ and $(d^2R/dt^2/)(t)$.

It is known in the Newtonian theory of gravitation that the gravitational force inside a uniform spherical shell of matter is zero. Thus if we imagine the whole universe divided up into thin spherical shells centred on O, the shells exterior to the shell ABCD defined above will have no gravitational influence on the particles A, B, C, D. According to O, the only force acting on the particles is the gravitational attraction of the matter interior to the shell ABCD (see Fig. 4.3). Thus by Newton's second law of motion, the force on a particle of mass m on this shell

$$m\underline{\ddot{r}} = - \frac{4 \pi G m \rho}{3} \underline{r}.$$

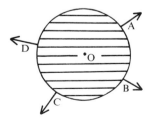

FIG. 4.3. According to O, the only force acting on A,B,C,D is the gravitational attraction of the shaded sphere, centred on O.

Substituting from eqn (4.8) and cancelling m,

$$\ddot{R}\,\frac{r_0}{R_0} = -\frac{4\pi\,G\rho R}{3}\frac{r_0}{R_0},$$

so

$$\ddot{R} = -4\pi\,G\rho R/3. \tag{4.11}$$

This can be integrated, using eqn (4.10), to give

$$\dot{R}^2 = \frac{8\pi G\rho_0 R_0^{\,3}}{3R} - k\,c^2, \tag{4.12}$$

where k is an arbitrary constant, dimensionless if R is taken
to have dimensions of distance.

   Before we look for solutions of this, we shall consider what
kinds of model are consistent with the cosmological principle in
Einstein's general theory of relativity.

4.4. THE SPECIAL AND GENERAL THEORIES OF RELATIVITY

   To explain these theories properly would need a whole book
in itself, so here I shall give only a thumbnail sketch.

   Newton postulated the existence of *inertial* frames of reference,
in which the motion of a free particle (i.e. with no forces
acting on it) would be a straight line. The surface of the earth
is quite a good approximation to an inertial frame, but there are
small deviations due to the earth's rotation. Once you have
found one inertial frame, all the others can be found by
performing one or both of the operations (a) changing the
origin, and rotating the coordinate axes; (b) moving with a
uniform velocity with respect to the original frame.

   Newtonian theory works very well for most purposes, but it
breaks down when we consider the velocity of light from moving
sources. It predicts that if we measure the velocity of light
from a source moving towards us, the result should be the sum
of the velocity of the source and the velocity of light itself.

However, the Michelson-Morley experiment showed that in fact the velocity of light is independent of the velocity of the source. Special relativity takes this fact as its starting point, and makes a modification, which under normal circumstances is small, to the way physical quantities change under the operation (b) above. The modification becomes large when relative velocities close to the velocity of light are involved. A moving body appears to have its length in the direction of motion contracted, its time slowed down, and its mass increased, all by a factor

$$\gamma = (1-v^2/c^2)^{-1/2},$$

where v is its velocity. This factor becomes infinite as v approaches c, so a body can never be accelerated past the speed of light. These effects have been accurately verified, in large particle accelerators, for example, as also has the famous prediction of the equivalence of mass and energy, summarized by $E = m c^2$.

So far we have dealt only with uniform relative velocities. We do not know how to deal with accelerated motions, and we no longer know what to do about gravity, since Newtonian theory is not compatible with special relativity. Einstein solved these problems with his general theory of relativity. This is based on the principle of equivalence, which stated most simply says that gravity disappears if you let yourself fall freely. Everything experiences the same acceleration under gravity (Galileo's experiment) so if you are falling too, the effects of gravity seem to vanish. At every point we can find a frame of reference in which special relativity holds locally (the local freely falling frame). Gravity is reduced to the status of a non-inertial force like centrifugal or coriolis force, which appears only because we have not chosen the right frame of reference to make our observations.

In relativity theory the idea of an *event*, something happening in a particular place at a particular time, plays a large role.

An event can be characterized by the four coordinates (x_1, x_2, x_3, t), where the first three are spatial coordinates and the fourth represents time. We can think of events as points in the four-dimensional *space-time* continuum. If a light signal is emitted at event (x_1, x_2, x_3, t) and received at a nearby event $(x_1 + dx_1, x_2 + dx_2, x_3 + dx_3, t + dt)$ (Fig. 4.4) then we know that, according to the special theory of relativity,

$$ds^2 = dt^2 - \frac{1}{c^2} (dx_1^2 + dx_2^2 + dx_3^2). \qquad (4.13)$$
$$= 0.$$

If the two events are not linked by a light signal, then $ds^2 \neq 0$.

FIG. 4.4. Two neighbouring events connected by a light signal.

The quantity ds is called the *interval* between the two neighbouring events, and in the special theory all inertial observers get the same answer for ds when they observe the same pair of events (this is obviously true when the two events are connected by a light signal, since the velocity of light is c in all inertial frames, so they all get the answer $ds = 0$). We say ds is an *invariant*.

In the general theory we require that ds be an invariant for *all* frames of reference, whether they are in uniform relative motion or not. In the particular case where we are in a freely falling frame, the expression for ds will simplify to eqn (4.13) and special relativity will hold.

But this elimination of the effect of gravity by choosing a freely falling frame only works locally. For bodies far enough away from us we notice that they are falling to the centre of the earth in a slightly different direction to ourselves. In general we will have to deal with *curved space-time* and the expression for the interval is

$$ds^2 = g_{11}dx_1^2 + g_{22}dx_2^2 + g_{33}dx_3^2 + g_{44}dx_4^2 + 2g_{23}dx_2dx_3 +$$
$$+ 2g_{24}dx_2dx_4 + 2g_{34}dx_3dx_4 + 2g_{12}dx_1dx_2 + 2g_{13}dx_1dx_3 +$$
$$+ 2g_{14}dx_1dx_4, \quad \text{where } x_4 = t,$$

or, more compactly,

$$ds^2 = \sum_{\lambda,\mu=1}^{4} g_{\lambda\mu} \, dx_\lambda \, dx_\mu, \tag{4.14}$$

where the $g_{\lambda\mu}$ are functions of position and time which determine the curvature of space-time and hence the gravitational field. Eqn (4.14) describes the *metric* of space-time, and the $g_{\lambda\mu}$ are called the components of the *metric tensor* ($g_{\lambda\mu}$ is chosen equal to $g_{\mu\lambda}$).

In a normal three-dimensional Euclidean space, the metric is simply

$$ds^2 = dx_1^2 + dx_2^2 + dx_3^2$$

and so $g_{11} = 1 = g_{22} = g_{33}$, and $g_{\lambda\mu} = 0$, $\lambda \neq \mu$. The metric of special relativity (called the Minkowski metric) is, by eqn (4.13),

$$ds^2 = -\frac{1}{c^2}(dx_1^2 + dx_2^2 + dx_3^2) + dx_4^2,$$

where $x_4 = t$, so

$$g_{11} = g_{22} = g_{33} = -\frac{1}{c^2} \; ; \; g_{44} = 1, \; g_{\lambda\mu} = 0, \lambda \neq \mu.$$

But in a general frame of reference the $g_{\lambda\mu}$ will vary with position and time, and the geometry will be that of a curved space-time. This is the mathematical way of describing the effect of gravity. Of course, the magnitude of this curvature is very small near the earth (only one part in 10^9).

In Newtonian theory and special relativity, the orbit of a free particle or a photon was a straight line. In the general

theory the role of straight line is taken by *geodesics*, the
shortest routes between pairs of points in a curved space. It
is then found that light passing near the sun is deflected
through a small angle or order $2GM_\odot/R_\odot c^2$ (in radians) where $M_\odot$
and $R_\odot$ are the mass and radius of the sun. Suppose now the sun
were compressed by a very large factor until this quantity became
greater than unity. Light near the sun would then be so strongly
deflected that in fact none could escape. The sun would disappear
from view, becoming a *black hole* (see p.27). For massive stars
it seems that this is indeed their probable fate (section 2.4).
Other effects of general relativity include the gravitational
red-shift, mentioned in section 3.3, and the advance of the
perihelia of the planets, most pronounced for Mercury.

4.5. GENERAL RELATIVISTIC COSMOLOGY

It can be shown that the most general metric satisfying the
cosmological principle is the *Robertson-Walker metric*:

$$ds^2 = dt^2 - \frac{R^2(t)}{c^2}\left(\frac{dr^2}{1 - k\,r^2} + r^2\,d\,\theta^2 + r^2\,\sin^2\theta\,d\,\phi^2\right),$$

$$(4.15)$$

where (r, θ, ϕ) are spherical polar coordinates, and r is chosen
for simplicity to be a *co-moving* radial coordinate, i.e. we are
picking a coordinate system in which fundamental observers
(see section 4.1 — we normally identify these with galaxies)
have the same radial coordinate all the time, even if the
universe expands or contracts. k is the curvature constant, and
if $k \neq 0$, it is convenient to redefine the units of the variable
r so that $k = \pm 1$. $R(t)$ is again the *scale factor* and we still
have the result that, as t changes, all spatial dimensions are
simply scaled up by the factor $R(t)$. Thus exactly the same
motions, isotropic expansion or contraction, are permitted in
general relativity as in Newtonian cosmology (section 4.3).

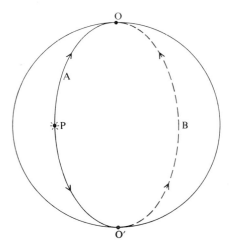

FIG. 4.5. An example of a two-dimensional space of positive curvature, the surface of a sphere. A signal confined to the surface and travelling by the shortest route will travel on a geodesic, in this case the great circle joining the source P and observer O. The signal has two alternative routes, PAO and PO´BO.

If we consider all events with the same value of t, so that $dt = 0$, the metric (4.15) can be shown to correspond to a three-dimensional space of constant curvature. If $k = +1$, we have a space of positive curvature (elliptic space — see Fig. 4.5); if $k = 0$ we have normal flat space; and if $k = -1$, we have a space of negative curvature (hyperbolic space).

To solve for $R(t)$ we must substitute this metric into the *field equations*, differential equations relating the metric functions $g_{\lambda\mu}$ to the density and pressure of matter — the general relativistic analogue of the Newtonian equations of motion for a fluid. When we do this we again obtain not only eqn (4.9), which is not too surprising since it expresses the conservation of mass, but also eqns (4.11) and (4.12).

At first sight it seems amazing that the general theory of relativity yields exactly the same cosmological models as Newtonian theory (the two theories do not yield the same results for the motion of the planets). But it is less strange when we remember (a) that the general theory of relativity is designed to reduce to Newton's law of gravitation when the gravitational field is very weak, e.g. in our cosmological neighbourhood, and (b) the cosmological principle requires that each neighbourhood be identical to every other.

Before we study the properties of eqn (4.12), we should note that in the original form of the field equations proposed by Einstein, an additional term appeared, the so-called 'cosmological term'. Later, Einstein argued that this term should be dropped, and this is the view of most relativists today. In Chapter 8 certain interesting consequences of the cosmological term are discussed, but for the moment it is set equal to zero.

4.6. CLASSIFICATION OF COSMOLOGICAL MODELS

(a) *The Milne model*, $\rho = 0$, $k = -1$

This is a universe of particles of negligible mass, so can also be called the special relativity cosmology. The solution of eqn (4.12) is

$$R(t) = \pm ct, \tag{4.16}$$

where we have chosen $t = 0$ to correspond to $R = 0$. The universe expands (or contracts) uniformly and monotonically (Fig. 4.6).

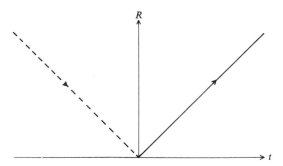

FIG. 4.6. The Milne (special relativity) model, $R \propto t$. The contracting solution is shown as a broken line.

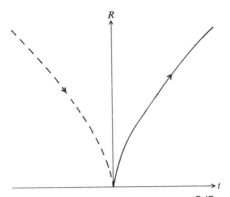

FIG. 4.7. The Einstein-de Sitter model, $R \propto t^{2/3}$. The contracting solution is shown as a broken line.

(b) *The Einstein-de Sitter model,* $k = 0$

The solution of eqn (4.12) is

$$R(t) = \pm R_0 \, (t/t_0)^{2/3} \qquad (4.17)$$

and the universe again expands monotonically, but at an ever-decreasing rate (see Fig. 4.7).

(c) $\rho > 0, k = -1$

$\dot{R}^2 > 0$ for all R, so R keeps changing monotonically (Fig. 4.8).

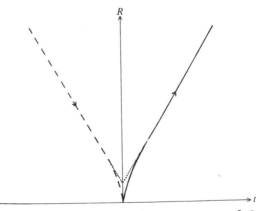

FIG. 4.8. The general $k = -1$ case. For small $t, R \propto t^{2/3}$, and for large $t, R \propto t$ (asymptote shown dotted). The universe expands (or contracts - broken line) monotonically.

As t gets very large, $\dot{R} \rightarrow \pm c$, so the universe looks more and more like a Milne model. As the galaxies get very far apart, their mutual gravitational attraction plays an ever weaker role in determining their motion.

(d) $k = +1$

$\dot{R}^2 = 0$ for a particular value of R,

$$R_c = \frac{8\pi G\rho_0 R_0^3}{3c^2}$$

and since $\ddot{R} < 0$ for all R, the expansion is halted (by the mutual gravitational attraction between the galaxies) and turned into a contraction. We have an oscillating universe (see Fig. 4.9).

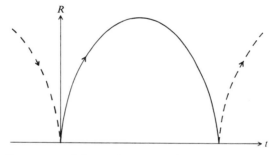

FIG. 4.9. The general k = +1 case. An oscillating universe. It is not known if the universe can go through more than one cycle.

Note that (i) whatever the value of k, the first term on the right-hand side of eqn (4.12) must dominate for small values of R, so the universe always looks like an Einstein-de Sitter model in the early stages, if $\rho > 0$, (ii) all these models are big-bang universes, in that $R \rightarrow 0$ at a finite time in the past (for expanding models), and the universe emerged from a 'singularity' (infinite density) at that time.

This follows very simply from eqn (4.11), which tells us that the $R(t)$ curve must be concave downwards and so must intersect the $R = 0$ axis at a finite time in the past. Note that this big bang is not an explosion into a pre-existing void. Because of the assumption of homogeneity, the whole universe is involved in the expansion and there is no 'outside' to expand into.

4.7. COSMOLOGICAL PARAMETERS

We have already encountered (section 3.3) the *Hubble parameter*

$$H(t) = \dot{R}(t)/R(t). \qquad (4.18)$$

We now define the *deceleration parameter*

$$q(t) = - R(t)\ddot{R}(t)/\dot{R}^2(t) \qquad (4.19)$$

and the *density parameter*

$$\sigma(t) = 4\pi G\rho(t)/3H^2(t). \qquad (4.20)$$

Eqn (4.11) implies that $\sigma(t) = q(t)$, at all values of t and (4.12) implies that

$$kc^2 = R^2H^2(\sigma q - 1). \qquad (4.21)$$

Thus whether the curvature constant k is +1, 0, or -1 is determined by whether q is greater than, equals, or is less than, 1/2. The only models in which q (and hence σ) does not change with time are the Milne ($\sigma = q = 0$) and Einstein-de Sitter ($\sigma = q = 1/2$) models.

The currently accepted values of these parameters at the present epoch $t = t_0$, are $\tau_0 = H_0^{-1} \simeq 2 \times 10^{10}$ years, with an uncertainty of no greater than a factor of 2 (see Fig. 3.1; p.51), $-1 \lesssim q_0 \lesssim 2$ (from the magnitude-red-shift test - see Chapter 6); and $0.01 \lesssim \sigma_0 \lesssim 0.1$ (from the observed density of matter in galaxies - see Chapter 7).

The value of the density appropriate to the Einstein-de Sitter model would be

$$\rho_{ES} = 3H_0^2/8\pi G = 4 \times 10^{-27} \text{kg m}^{-3}; \qquad (4.22)$$

this is often referred to as the *critical density*, since if

$\rho_0 > \rho_{ES}$ we are in an oscillating universe and if $\rho_0 \lesssim \rho_{ES}$ we are in a monotonic expanding one.

4.8. THE AGE OF THE UNIVERSE

We can write

$$t_0 = \int_0^{t_0} dt = \int_0^{R_0} \frac{dR}{\dot{R}}$$

and using eqns (4.12) and (4.18)-(4.21):

$$t_0 = \tau_0 \int_0^1 \frac{dx}{(2q_0/x + 1 - 2q_0)^{1/2}} , \qquad (4.23)$$

where we have written x for R/R_0. If $q_0 = 0$, $t_0 = \tau_0$; if $q_0 = 1/2$, $t_0 = 2\tau_0/3$.

For $0 < q_0 < 1/2$, and $q_0 > 1/2$, the substitutions

$$x = \frac{2q_0}{1 - 2q_0} \sinh^2\theta, \text{ and } x = \frac{2q_0}{2q_0 - 1} \sin^2\theta, \qquad (4.24)$$

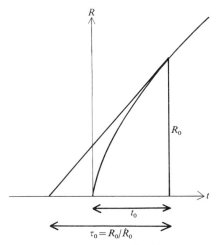

FIG. 4.10. The Hubble time τ_0 exceeds the age of the universe t_0, if $\ddot{R} \lessgtr 0$ for all t.

respectively, allow eqn (4.23) to be evaluated. The age of the universe is less than or equal to the Hubble time in all models: the reason for this is illustrated in Fig. 4.10.

PROBLEMS

4.1 Evaluate eqn (4.23) using eqns (4.24) (see Fig. 4.11).

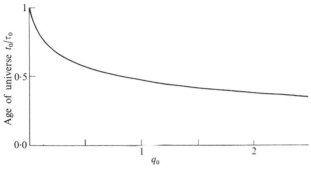

FIG. 4.11. Dependence of the age of the universe, as a fraction of the Hubble time, on the present-day deceleration parameter q_0.

4.2. Show that q is a constant in the Milne and Einstein-de Sitter models.

4.3. Show that eqn (4.12) has solutions of the form

$$R(t) = \frac{c\tau_0 q_0}{(2q_0 - 1)} 3/2 (1 - \cos 2\psi), \quad t = \frac{q_0 \tau_0}{(2q_0 - 1)} 3/2 (2\psi - \sin 2\psi),$$

if $k = + 1$

and $R(t) = \frac{c\tau_0 q_0}{(1 - 2q_0)} 3/2 (\cosh 2\psi - 1), \quad t = \frac{q_0 \tau_0}{(1 - 2q_0)} 3/2 (\sin 2\psi - 2\psi),$

if $k = -1$,

where ψ ranges over all real values.

5. Early stages of the big-bang

5.1. UNIVERSE WITH MATTER AND RADIATION

So far we have constructed models of universes filled only with matter. Although radiation contributes less than 1 per cent of average energy per unit volume in the universe at the present epoch, it plays a crucial and dominant role in the early stages.

We can study the evolution of a universe containing matter and radiation by applying the first law of thermodynamics to an element of the substratum. This gives the same answer as applying the full field equations of general relativity.

The first law of thermodynamics states that the change in energy of an expanding system equals the work done by the pressure

$$dE = - p \, dV, \tag{5.1}$$

where E, p, V are the energy, pressure, and volume of the element.

We now use Einstein's equation

$$E = Mc^2, \tag{5.2}$$

where M includes both the contribution of the matter and the mass equivalent of the radiant energy, i.e.

$$E = (\rho_m + \rho_r)Vc^2 = \rho Vc^2, \tag{5.3}$$

where ρ_m is the density of the matter, ρ_r is the mass density of the radiation ($= U/c^2$, where U is the energy density), and ρ is the total mass density of matter and radiation. Since the volume of an element of the substratum is proportional to $R^3(t)$ we have, from eqn (5.1),

$$\frac{d}{dt}(\rho R^3) + \frac{p}{c^2} \frac{d}{dt} (R^3) = 0. \tag{5.4}$$

Universe with matter only

If the pressure of the matter can be neglected, eqn (5.4) reduces to

$$\frac{d}{dt}\,(\rho_m R^3) = 0,$$

as before (eqn (4.9)).

Universe with radiation only

It can be shown that the relation between the pressure and density for radiation is

$$p_r = \rho_r c^2/3. \tag{5.5}$$

Thus from eqn (5.4),

$$\frac{d}{dt}(\rho_r R^3) + \frac{1}{3}\,\rho_r\,\frac{d}{dt}(R^3) = 0,$$

or

$$3\rho_m R^2 \dot{R} + \dot{\rho}_r R^3 + \rho_r R^2 \dot{R} = 0$$

so

$$\frac{d}{dt}(\rho_r R^4) = 0. \tag{5.6}$$

This integrates to

$$\rho_r = \rho_{r,0}\left(R(t)/R_0\right)^{-4} \tag{5.7}$$

When this is substituted in eqn (4.12), assuming $k = 0$, we find $\dot{R}^2 \propto R^{-2}$, $\dot{R} \propto R^{-1}$, which integrates to

$$R(t) \propto t^{1/2}. \tag{5.8}$$

This characterizes the motion of a radiation-dominated universe in its early stages, since the term in eqn (4.12) in kc^2 becomes negligible if R is sufficiently small.

Universe containing matter and radiation

If we neglect the contribution of matter to the pressure, so that $p = p_r$, then eqn (5.4) becomes

$$\frac{d}{dt}(\rho_m R^3) + \frac{1}{R}\frac{d}{dt}(\rho_r R^4) = 0. \tag{5.9}$$

If we assume strict conservation of matter, i.e. we neglect any conversion of matter to radiation, then each of the two terms in eqn (5.9) will be separately zero:

$$\frac{d}{dt}(\rho_m R^3) = 0, \quad \frac{d}{dt}(\rho_r R^4) = 0$$

and so

$$\rho_m = \rho_{m,0}(R/R_0)^{-3}, \quad \rho_r = \rho_{r,0}(R/R_0)^{-4}. \tag{5.10}$$

However small the current value of the ratio of the density of radiation to matter $\rho_{r,0}/\rho_{m,0}$ (current value $\sim 10^{-3}$), there was an epoch in the past, given by

$$R_{crit} = \frac{\rho_{r,0}}{\rho_{m,0}} \cdot R_0 \tag{5.11}$$

such that $\rho_r > \rho_m$ for $R < R_{crit}$, and $\rho_r \gg \rho_m$ for $R \ll R_{crit}$. Radiation would have been the dominant form of energy at early epochs, and eqn (5.8) would have been valid.

We call epochs such that $\rho_r > \rho_m$ the *radiation-dominated era* and epochs such that $\rho_m > \rho_r$ the *matter-dominated era*.

5.2. THE FIREBALL

At the present epoch radiation traverses the universe freely, with only a small probability of being scattered by gas or dust. In addition to occurring in galaxies, such material may be spread more or less uniformly throughout space, but so tenuously that a photon is likely to travel several Hubble distances (see p.50.) before being scattered or absorbed. We say that the universe is transparent or *optically thin* at the present epoch. It is easy to see that this was not always so. The typical size of a

galaxy is 10 kpc, while the average spacing between galaxies is
of order 1 Mpc. If we run the universe backwards through a change
Z in the scale factor of 100, without galaxies altering, i.e.

$$Z = R(t_0)/R(t) = 100,$$

then the galaxies would all have been touching. This makes it
likely that for $Z > 100$ the universe presented a rather uniform
appearance, the break-up into galaxies occuring during epochs
such that $100 \gtrsim Z > 1$. Of course, galaxies would not remain
unaltered during this 'rewind' of the universe. We would see
the dust falling back into the stars from which it was blown
out, and the stars dissolving into the gas clouds from which
they formed. The atoms from which our complex chemistry derives
would have broken down into hydrogen with an admixture of helium.
We are led to a picture of a fairly uniformly distributed gas of
hydrogen and helium, spotted with the density irregularities which
are to become the galaxies.

All the while that we are running the universe backwards, the
energy density, and hence the temperature, of the radiation is
building up, by eqn (5.10). Sooner or later the gas will start
to be significantly heated by the radiation. The crucial moment
comes when the temperature of the matter reaches about 3000 K, for
then the hydrogen starts to become ionized. This brings into
play the enormous scattering power of free electrons and puts an
end to the transparency of the universe to radiation. We call
this moment the epoch of *decoupling* of radiation and matter.
Prior to this, they are locked together in thermal equilibrium.
This means that the radiation has the Planck black-body spectrum
(section 1.4; p.12), so the total energy density of the radiation

$$\rho_r c^2 = \int_0^\infty u_r (\nu) c^2 d\nu,$$

where the specific energy density is $u_r(\nu) = 4\pi I(\nu)/c$, and $I(\nu)$
is the intensity (eqn 1.2).

$$\rho_r c^2 \propto \int_0^\infty \frac{\nu^3 \, d\nu}{\exp(h\nu/kT_r) - 1} \propto T_r^4 \,, \tag{5.12}$$

where T_r is the radiation temperature, and then eqn (5.10) implies that

$$T_r \propto 1/R(t). \tag{5.13}$$

As $R \to 0$, $T_r \to \infty$, which explains the use of the term *fireball* to describe this optically thick phase of a big bang universe. Since there is thermal equilibrium prior to the epoch of decoupling, the temperature of the matter is the same as that of the radiation:

$$T_m = T_r. \tag{5.14}$$

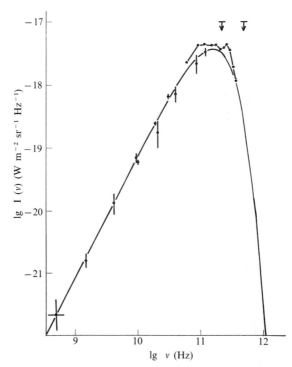

FIG. 5.1. A summary of observations of the intensity of the microwave background radiation as a function of frequency, compared with a 2.7 K black body.

What happens after the epoch of decoupling ? Certainly eqn
(5.14) will no longer hold and the matter will cool off rapidly.
However, it turns out that the effect of the expansion of the
universe on the radiation is to preserve its black-body spectrum,
with the radiation temperature continuing to fall according to
eqn (5.13). This provides the most natural explanation of the
2·7 K black-body microwave background radiation described in
section 1.8. A more detailed summary of the observations is
shown in Fig. 5.1. The discovery of this radiation provided the
most spectacular confirmation to date of the hot big-bang picture
of the universe.

Now, from eqn (5.13) we can identify the epoch of decoupling
as

$$Z = R(t_0)/R(t) = 3000/2\cdot 7 \sim 1000.$$

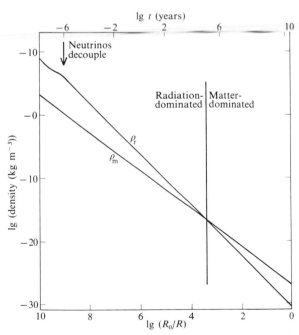

FIG. 5.2. The variation of the density of matter and radiation
with epoch.

The density of matter at this time would have been

$$\rho_m \sim (10^3)^3 \rho_{m,0} \sim 10^{-18} \text{kg m}^{-3}, \text{ assuming } \rho_{m,0} \sim 10^{-27} \text{ kg m}^{-3},$$

about 1000 times the average density of matter inside our Galaxy. Thus the universe was probably very uniform at this time.

Coincidentally, the critical epoch separating the radiation-dominated from the matter-dominated eras (section 5.1; p.84) is also about $Z \sim 1000$ for $\rho_{m,0} \sim 10^{-27} \text{ kg m}^{-3}$. Hence for $Z > 1000$ we can use $R(t) \propto t^{1/2}$ (eqn (5.8)), and for $1000 > Z \gg 1$ we can use $R(t) \propto t^{2/3}$ (eqn (4.17)). In Figs. 5.2 and 5.3, which show the variation of the energy densities and temperatures of matter and radiation with epoch, we use the latter right up to $Z = 1$, for illustration.

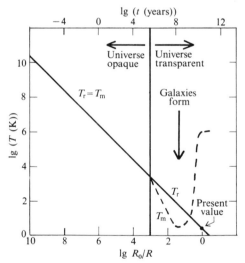

FIG. 5.3. The variation of the temperature of matter and radiation with epoch.

5.3. HELIUM PRODUCTION

When the temperature was between 10^{10} K and 10^9 K, some proportion of the hydrogen was converted by thermonuclear fusion into helium, the exact amount depending on the density of the

TABLE 4

Relative cosmic abundances
of the most common elements

(a) Helium

Location; method of evaluation	Percentage Helium by mass
Initial value in sun, from stellar evolution calculations	27
Solar cosmic rays	20-26
Globular cluster stars (apart from some anomalous stars with no visible helium)	26-32
Planetary nebulae	40
H_{II} regions	27
Best value	27

(b) The most common elements (abundance by mass relative to hydrogen)

Hydrogen	1
Helium	0.38
Carbon	4.0×10^{-3}
Nitrogen	1.3×10^{-3}
Oxygen	1.05×10^{-2}
Neon	1.7×10^{-3}
Magnesium	6.4×10^{-4}
Silicon	9.4×10^{-4}
Sulphur	5.1×10^{-4}
Argon	2.5×10^{-4}
Iron	2.25×10^{-3}
Nickel	1.2×10^{-4}

matter during this phase. For $\rho_{m,0} \sim 10^{-27}$ kg m^{-3},

$$\rho_m(T = 10^{10} \text{ K}) \sim 10^{+1 \cdot 8} \text{kgm}^{-3},$$

and it is found that the fraction of matter converted to helium would be almost exactly the 27 per cent by mass that we need to explain the composition of our Galaxy (Fig. 5.4). This provides a second major success for big-bang cosmology.

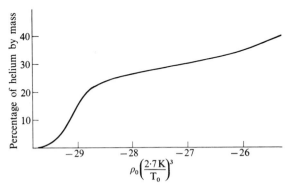

FIG. 5.4. Percentage of helium by mass formed in the fireball, as a function of the present density of the Universe.

We have pushed the fireball scenario back to within 1 second of the big bang. Extrapolating what is known about elementary particles, we can get even closer, but we should bear in mind that even the most optimistic interpretation of the $0 \cdot 1$ per cent isotropy of the black-body radiation guarantees our simple isotropic models only back to $Z \sim 10^6$, so we are treading on thin ice.

We can divide the early fireball into four regimes:

(a) $T > 10^{12}$ K: there are a great variety of particles in thermal equilibrium with each other, including photons, leptons, mesons, and nucleons, and their antiparticles.

(b) $T < 10^{12}$ K: muons annihilate; neutrinos and antineutrinos decouple from everything else.

(c) $T \sim 5 \times 10^9$ K: electron-positron pairs annihilate, leaving only photons, neutrinos, and antineutrinos, and a small proportion of protons, neutrons, and electrons.

(d) $T \sim 10^9$ K: neutrons and protons fuse into heavier elements (some of the reactions are shown below), creating He^4 and a trace of ^2H (deuterium), ^3He, ^7Li and other elements (see Fig. 5.5).

$$n + p \longleftrightarrow {}^2H + \gamma$$

$$^2H + {}^2H \longleftrightarrow {}^3He + n$$

$$^3He + n \longleftrightarrow {}^3H + p$$

$$^3H + {}^2H \longleftrightarrow {}^4He + n$$

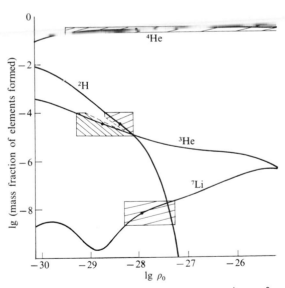

FIG. 5.5. The mass fraction of the elements ^4He, ^3He, ^2H (deuterium), and ^7Li formed in the fireball, as a function of the present density of the universe, compared with observed primordial values (shaded boxes). A density not far from 10^{-28} kg m^{-3} seems to be required.

PROBLEM

5.1. Show that the substitution $\nu' = \nu/Z$ into the expression for the intensity of black-body radiation $I(\nu)$ given by eqn (1.2) results in a black-body spectrum corresponding to temperature $T' = T/Z$. Give an interpretation of this result.

6. Observational cosmology

6.1. INTRODUCTION

The isotropic 2·7 K black-body radiation, the 27 per cent cosmic helium abundance, and the similarity between the ages of galaxies and the age of the universe all support the big-bang models derived from general relativity assuming the cosmological principle. Can we test these models in more detail? In particular, can we determine the current value of the deceleration parameter q (section 4.7) and thereby deduce what the future of the universe will be? In this chapter we look at a variety of cosmological tests, all using discrete sources of radiation, that have been applied to try to answer this question. The answer turns out to be inextricably bound up with the *evolution* of the different classes of discrete source in the universe.

The main tests involve comparing luminosity and diameter distance (section 3.2) with red-shift, source counts (sections 1.7 and 3.5), and integrated background radiation (section 1.8). It was radio-source counts that first showed that strong evolution must affect some populations of source.

6.2. NEWTONIAN THEORY

Suppose that we are in an expanding Newtonian universe, in a Euclidean geometry. The light from a source at distance d, receding with velocity $v = H_0 d$ (where H_0 is the Hubble constant — section 3.3), will suffer a Doppler shift

$$\frac{\nu_e}{\nu_o} = \frac{\lambda_o}{\lambda_e} = 1 + \frac{\Delta\lambda}{\lambda_e} = 1 + z = Z, \qquad (6.1)$$

where λ_e, ν_e, and λ_o, ν_o are the wavelengths and frequencies of emission and observation, respectively, and $\Delta\lambda = \lambda_o - \lambda_e$, of magnitude

$$z = v/c. \tag{6.2}$$

The flux S from the source is related to its luminosity P by the inverse-square law

$$S = P/d^2, \tag{6.3}$$

and the apparent angular size θ (rad) of an object of linear size l is given by

$$\theta = l/d. \tag{6.4}$$

The number of sources per steradian with distances $< d$ is $\eta\, d^3/3$, where η is the number of sources per unit volume (the *number density*), and if all the sources have the same luminosity P, then the number of sources per steradian with fluxes brighter than S,

$$N(S) = \frac{1}{3}\,\eta\,(P/S)^{3/2}, \tag{6.5}$$

The relations (6.2)-(6.5) might be expected to hold generally for $v \ll c$, i.e. $z \ll 1$, but we know they will break down as $v \to c$ because of the effects of special relativity.

6.3. SPECIAL RELATIVITY COSMOLOGY: THE MILNE MODEL

We now take into account the effects of special relativity, but neglect the effects of gravitation. If we consider particles moving with the substratum, observing them from an inertial frame, then there are no forces acting on them so they all move with uniform velocity with respect to each other. A fundamental observer sitting on one of these particles continues to use Euclidean space coordinates, and measures the velocity of a particle at position vector $\underline{r}$ at time t to be $\underline{v}(\underline{r},t)$. The only motion consistent with the cosmological principle turns out to be

$$\underline{v}(\underline{r},t) = \underline{r}/t \tag{6.6}$$

(from eqn (4.7); this corresponds to taking $R(t) \propto t$, as we expect from section 4.6; p.76). All particles would have been at the origin at $t = 0$ and then expand out isotropically with uniform velocity.

The Doppler shift can be shown to be

$$\frac{\nu_e}{\nu_0} = z = \frac{1 + v/c}{(1 - v^2/c^2)^{1/2}} \tag{6.7}$$

(which agrees with the Newtonian expression (6.2) provided $v/c \ll 1$), and the flux from a distant source is now given by

$$S = (P/r^2)z^{-4}, \tag{6.8}$$

where r is the distance of the source at the moment of emission, the factor z^{-4} taking account of the various special relativistic effects of the souce's motion on its apparent brightness. If the signal is received at time t_0, it was emitted at time $t_0 - r/c$, and eqn (6.6) implies that

$$r = v(t_0 - r/c), \text{ or } r = vt_0/(1 + v/c).$$

Eqn (6.8) becomes

$$S = \frac{P}{ct_0(z + z^2/2)^2}, \tag{6.9}$$

using eqn (6.7). The number of sources per steradian, with red-shift factors less than or equal to z is found to be

$$N(z) = \eta_0(ct_0)^3(z^2/8 - 1/8z^2 - \frac{1}{2}\ln z), \tag{6.10}$$

where η_0 is the local number density of souces at the present epoch.

Although the Milne model could be a reasonable approximation at the present epoch (we saw in section 4.6 (p.76) that all general relativity models with $k = -1$ tend to the Milne model for large t), it cannot be valid back to $t = 0$, since the density of matter becomes infinite as $t \to 0$ so that gravitational effects cannot be neglected.

6.4. GENERAL RELATIVISTIC COSMOLOGY: THE RED-SHIFT

Our starting point is the Robertson-Walker metric (section 4.5):

$$ds^2 = dt^2 - \frac{R^2(t)}{c^2} \left(\frac{dr^2}{1 - kr^2} + r^2 d\theta^2 + r^2 \sin^2\theta d\phi^2 \right), \quad (6.11)$$

together with the fact that for two events (r, θ, ϕ, t), $(r + dr, \theta + d\theta, \phi + d\phi, t + dt)$ connected by a light signal, the interval

$$ds = 0 \quad \text{(section 4.4)}.$$

Consider a photon emitted by a source at Q at time t_e, so that the event of emission is $(r_0, \theta_0, \phi_0, t_e)$ and suppose that the photon is received by an observer at the origin at time t_0, so that the event of observation is $(0, \theta_0, \phi_0, t_0)$ (see Fig. 6.1).

Source, Q
$r = r_0$
$t = t_e$

Observer, O
$r = 0$
$t = t_0$

FIG. 6.1. The radial null geodesic linking the events of emission and reception of a signal sent from Q to O.

Clearly the light signal travels in a radial straight line, by symmetry, so $d\theta = d\phi = 0$ and

$$ds^2 = 0 = dt^2 - \frac{R^2(t)dr^2}{c^2(1 - kr^2)}$$

for any element of the light ray joining Q to O, or

$$\frac{dr}{(1 - kr^2)^{\frac{1}{2}}} = - \frac{cdt}{R(t)} \quad (6.12)$$

for an incoming signal. Integrating this from Q to O

$$\int_0^{r_0} \frac{dr}{(1 - kr^2)^{\frac{1}{2}}} = \int_{t_e}^{t_0} \frac{cdt}{R(t)}. \quad (6.13)$$

Now recall that we chose r to be a *co-moving* coordinate (section 4.5), so that at a later time $t_o + dt_o$ the source is still defined by $r = r_0$ (its change in distance is incorporated into the scale factor $R(t)$). The left-hand side of eqn (6.13) therefore does not change with time for a particular source-observer pair. Now consider a second signal emitted by Q at time $t_e + dt_e$, and suppose this later signal is received by O at $t_o + dt_o$. Then eqn (6.13) becomes

$$\int_0^{r_0} \frac{dr}{(1 - kr^2)^{\frac{1}{2}}} = \int_{t_e + dt_e}^{t_0 + dt_o} \frac{cdt}{R(t)} \qquad (6.14)$$

for this later signal. Subtracting eqn (6.14) from (6.13),

$$\int_{t_e}^{t_e + dt_e} \frac{cdt}{R(t)} = \int_{t_0}^{t_0 + dt_0} \frac{cdt}{R(t)} ,$$

or

$$\frac{dt_e}{R(t_e)} = \frac{dt_o}{R(t_o)} \qquad (6.15)$$

(assuming dt_e/t_e, $dt_o/t_o \ll 1$). Now suppose the two events of emission correspond to consecutive wave crests (Fig. 6.2). Then

$$\frac{\nu_e}{\nu_o} = \frac{dt_o}{dt_e} = \frac{R(t_o)}{R(t_e)} = 1 + z. \qquad (6.16)$$

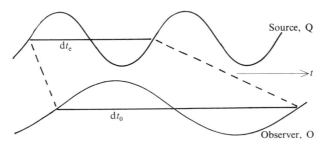

FIG. 6.2.   The red-shifting of a light-wave from a distant source.

If the universe has expanded so that $R(t_o) > R(t_e)$, then there is a red-shift ($z > 0$)

If we could observe light which was emitted from a source at $t_e = 0$, so $R(t_e) = 0$, it would be red-shifted to infinite wavelength. However, as the universe is opaque for $R(t)/R(t_0) < 10^{-3}$ (section 5.2), we know that we can observe no sources with red-shift greater than 1000.

6.5. LUMINOSITY DISTANCE

To calculate the flux from a distant source Q, consider a spherical surface $r = r_o$ centred on Q, passing through the observer O. Then the element of area at O defined by the four points (θ,ϕ), $(\theta + d\theta,\phi)$, $(\theta,\phi + d\phi)$, $(\theta + d\theta, \phi + d\phi)$, will subtend a solid angle

$$d\Omega = \sin \theta d\theta d\phi$$

at Q (see Fig. 6.3). To calculate the area of this element we

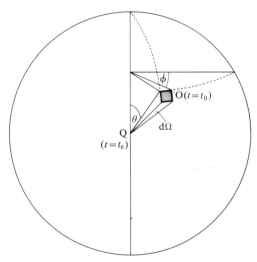

FIG. 6.3. The light emitted into a small element of solid angle $d\Omega$ is received by O on the small shaded element of area, which forms part of the sphere $r = r_0$ centred on Q.

note that the 'proper' distance (i.e. as determined by radar methods) between two events is given by

$$(-c^2 ds^2)^{1/2} \big|_{dt = 0}$$

so the area of the element =

$$R(t_o)r_o d\theta \quad . \quad R(t_o)r_o \sin \theta d\phi = R_o^2 r_o^2 d\Omega. \quad (6.17)$$

For a unit area,

$$d\Omega = (R_o^2 r_o^2)^{-1}$$

The energy emitted per second into $d\Omega$ is $Pd\Omega$, so the flux received by O per unit area

$$S = \frac{P}{R_o^2 r_o^2} z^{-2}, \quad (6.18)$$

where one factor, z^{-1}, is needed because the photons arrive with less energy than they set out with (since $E = h\nu$) and the second factor, z^{-1}, because the photons arrive less frequently than they set off by eqn (6.15).

From the definition of luminosity-distance (eqn (3.1)), eqn (6.18) implies

$$D_{lum} = R_o r_o z \quad (6.19)$$

where r_o is related to the red-shift by eqn (6.13). Expanding eqn (6.19) in a power series in z,

$$D_{lum} = c\tau_o \{z - \frac{1}{2}(q_o - 1)z^2 + \cdots \}. \quad (6.20)$$

To the first order in z we simply have the Hubble law, but the second-order term depends term depends on the deceleration parameter q_o, which we can therefore hope to determine by observing $D_{lum}(z)$.

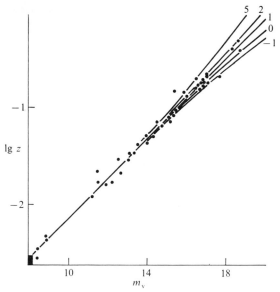

FIG. 6.4. Red-shift-magnitude curves for cosmological models with different values of q_o, compared with Sandage's data for the brightest galaxies in clusters. The small shaded rectangle in the lower left-hand corner indicates the maximum range of Hubble's 1927 observations. The curve labelled -1 corresponds to the prediction of the steady state cosmology.

In terms of magnitudes (section 3.2),

$$m = M + 5\lg(D_{\text{lum}}(z)/10\text{pc})$$

$$= M + 5\lg(c\tau_o/10\text{pc}) + 5\lg z - \frac{2 \cdot 5}{\ln 10}(q_o - 1)z$$

$$(6.21)$$

Exact curves for different values of q_o are shown in Fig. 6.4 compared with the observed magnitude-red-shift relation for the brightest galaxies in 40 clusters. The best value of q_o is

$$q_o = 1 \cdot 0 \pm 0 \cdot 4. \tag{6.22}$$

However, the true uncertainty is far larger since galaxies are probably changing their luminosity with time. If galaxies were more luminous in the past (due to a more rapid formation rate

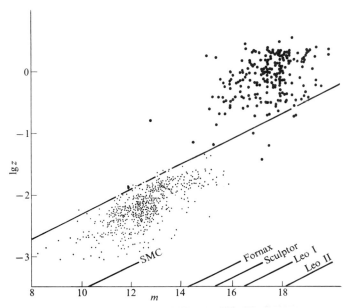

FIG. 6.5. The red-shift-magnitude relation for quasars (large dots, visual magnitude) and for all galaxies in the Reference Catalogue of Bright Galaxies (small dots, blue magnitude, Virgo cluster excluded). The solid line corresponds to the $q_o = 1$ line of Fig. 6.5: few galaxies more luminous than those brightest galaxies in clusters are found. The Catalogue (see Table 5. p.106) is complete only to 13th magnitude, and galaxies similar to the lowest luminosity objects in the Local Group could not be detected (where they would lie is shown). The luminosity of galaxies ranges over at least a factor of 100 000 and of quasars at least 10 000 (assuming their red-shifts are cosmological).

of bright stars, for example) then a smaller value of q_o is appropriate, and vice versa. Probably q_o lies between -1 and 2.

The scatter of the points in Fig. 6.4 about the mean line arises because these galaxies do not have exactly the same luminosity. If we plot *all* galaxies in such a diagram (Fig. 6.5). the scatter is enormous, since the absolute magnitude of galaxies ranges over at least 12 magnitudes (a factor of 100,000 in the optical luminosity P). The same is true for quasars, assuming their red-shifts are cosmological. Clearly we can not use Fig. 6.5 to determine q_o.

However, another class of objects that has been used to apply this test is the radio galaxies. By taking deep photographic plates at the positions of radio sources, some very faint galaxies have been found, some of which have red-shifts appreciably larger than the most distant cluster in Fig. 6.4. These will ultimately give useful information on q_o.

6.6. DIAMETER DISTANCE

Consider an object of size l at distance $r = r_o$, subtending an angle $\delta\theta$ at the origin (Fig. 6.6). From the metric (eqn

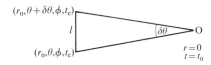

FIG. 6.6. An object of size l subtending an angle $\delta\theta$ at the observer O.

(6.11)), the proper distance between the ends of the object is

$$R(t_e)r_o\delta\theta \tag{6.23}$$

= l, by definition, so

$$\delta\theta = \frac{lz}{R_o r_o} \tag{6.24}$$

From eqn (3.5), the diameter distance is then

$$D_{diam} = R_o r_o z^{-1}$$

This has been applied to bright galaxies in clusters (Fig. 6.7); the quoted value of q_o is $0\cdot15 \pm 0\cdot3$. It is also hoped that the test can be applied to rich clusters of galaxies, which seem to have a core of well-defined linear size.

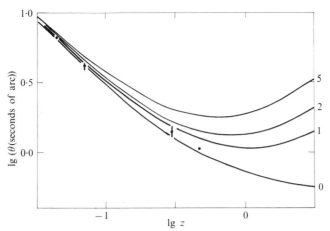

FIG. 6.7.   Angular diameter-red-shift diagram for galaxies in clusters.

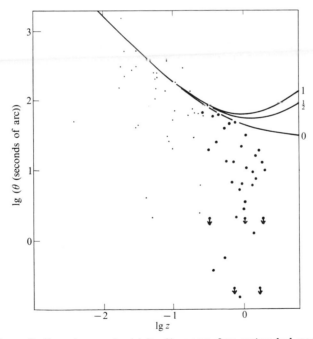

FIG. 6.8.   Radio-size-red-shift diagram for extended sources associated with radio galaxies (small dots) and quasars (large dots), showing the continuity between the two populations and the existence of an apparent upper envelope.

Notice that all the theoretical curves, except that for
$q_o = 0$, go through a minimum in $\delta\theta$, after which $\delta\theta$ starts to
increase with red-shift. It would be an important test of these
models to actually see this happening. Since the minimum occurs
at fairly large red-shift, unless q_o is unreasonably large, the
best hope for testing this lies in the quasars. Fig. 6.8 shows
a plot of the radio size of quasars with extended radio components
against red-shift. Some radio galaxies are also plotted to
illustrate the continuity between these two populations. Although
there is a great deal of scatter, there seems to be a well-
defined upper envelope. However, it does not seem to follow the
theoretical curves predicted from eqn (6.24), and this may have
to be interpreted as implying that extended radio sources
associated with quasars were smaller in the past.

6.7. NUMBER COUNTS OF SOURCES

Consider a population of sources uniformly and randomly
distributed through the matter in the universe, like the
currants in a pudding. The number density of sources

$$\eta(t) \propto \rho(t) \tag{6.25}$$

provided the probability of a piece of matter being a source is
independent of t. The proper volume of the element at Q is
(Fig. 6.9).

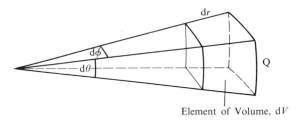

Element of Volume, dV

FIG. 6.9.   A small element of volume.

$$dV = \frac{R\ dr}{(1 - kr^2)^{1/2}} \cdot Rrd\theta \cdot Rr\ sin\theta\ d\phi \qquad (6.26)$$

and for 1 sr of a spherical shell

$$dV = \frac{R^3(t)r^2 dr}{(1 - kr^2)^{1/2}} \cdot$$

The number of sources in 1 sr of this shell is then

$$\frac{n(t)R^3(t)r^2 dr}{(1 - kr^2)^{1/2}} = \frac{\eta_o R_o^3 r^2 dr}{(1 - kr^2)^{1/2}}$$

by eqns (4.10) and (6.25), where $\eta_0 = \dot{n}(t_0)$.

The total number of sources per steradian out to $r = r_o$,

$$N(r_o) = \eta_o R_o^3 \int_0^{r_0} \frac{r^2\ dr}{(1 - kr^2)^{1/2}} \cdot \qquad (6.27)$$

This can then be combined with eqn (6.18) to give $N(S)$, the number of sources per steradian that are brighter than S, assuming all sources have the same luminosity.

It is found that

$$\frac{d\lg N}{d\lg S} > -1\cdot 5 \text{ for all } S, q_o.$$

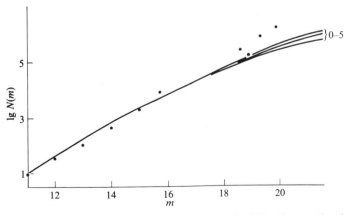

FIG. 6.10.   Counts of galaxies, compared with theoretical curves for different q_o.

TABLE 5

Some of the great catalogues of modern astronomy

OPTICAL

1. *The reference catalogue of bright galaxies*

Compiled by G. and A. de Vaulouleurs, this comprises essentially the galaxies from the New General and Index Catalogues (NGC† and IC†) of nebulae, in turn based on the General Catalogue of William and John Herschel. It contains information on 2599 galaxies, and is complete to about 13th magnitude.

2. *The Zwicky catalogue of galaxies and clusters of galaxies*

Compiled by F. Zwicky and co-workers, it contains 31 000 galaxies in the northern hemisphere brighter than $m_{pg} = 15 \cdot 7$, and lists 9700 clusters of galaxies.

3. *The Abell catalogue of rich clusters of galaxies*

Lists 2700 clusters rich in galaxies

4. Catalogues of unusual objects

(a) *Zwicky lists of compact galaxies* (b) *Arp catalogue of peculiar galaxies* (c) *Vorontsov-Velyaminov catalogue of interacting galaxies* (d) *Markarian catalogue of blue galaxies.*

RADIO†

Revised 3rd Cambridge Catalogue of northern hemisphere sources brighter than 9 Jansky $\{1 \text{ Jy} = 10^{-26} \text{ Wm}^{-2} \text{ Hz}^{-1}\}$ *at 178 MHz.*

Parkes Catalogue of southern hemisphere sources.

Bologna Catalogue of northern hemisphere sources brighter than 0.1 Jy at 408 MHz.

X-RAY†

The 3rd *Uhuru Catalogue* of bright sources in the range 2-8 keV, compiled from data from the Uhuru satellite.

---

†contains some Galactic sources also.

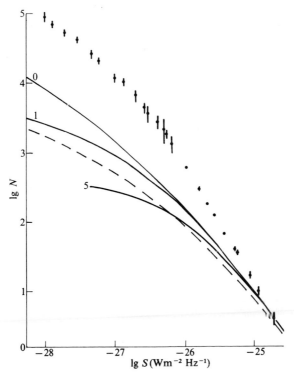

FIG. 6.11. Counts of radio sources, compared with theoretical curves for different q_o (solid curves) and for the steady-state model (broken curve — see section 8.3), showing strong evolutionary effects in the radio-source population.

Fig. 6.10 shows counts of galaxies. The observational uncertainties make the determination of q_o impossible.

Counts of radio sources (Fig. 6.11) are appreciably steeper than any of the theoretical curves, implying that strong evolutionary effects must be present. The sources (mostly quasars and radio galaxies) must either have been more luminous in the past, or the probability of a source being alight must have been greater.

6.8. THE LUMINOSITY-VOLUME TEST

These evolutionary effects can also be seen vividly by means of the *luminosity-volume* test, which combines the luminosity-distance and source-count tests into a single, more powerful test, provided a complete sample of sources down to some limiting flux level S_{min} is available. For a source in the sample with flux S and red-shift z we can calculate the luminosity P from eqn (6.18) for some chosen cosmological model. Sources of this

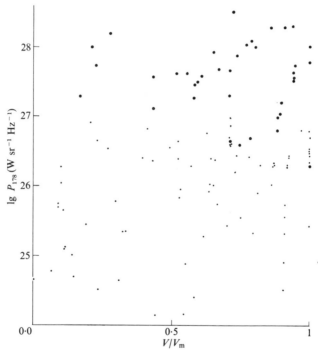

FIG. 6.12. The luminosity-volume test. The distribution of monochromatic radio luminosity (at 178MHz) against V/V_m for all quasars (large dots) and radio galaxies (small dots) in the revised 3rd Cambridge catalogue of radio sources (see Table 5 p.106) with visual magnitude brighter than 19.5, and with $|b| > 10^o$, $\delta > 10^o$, where b is the Galactic latitude and δ is the declination. The distribution for quasars and for the more luminous radio galaxies is non-uniform, showing that these populations have changed their properties dramatically with time. The calculations have been performed in the Milne model ($q_o = 0$).

luminosity P should then be uniformly distributed with respect to co-moving volume, $V(z)$ (i.e. when the effect of the expansion of the universe is allowed for). Of course, there will be a red-shift z_{max} at which a source of this luminosity P would disappear out of the sample (its flux would drop below S_{min} for $z > z_{max}$), so in fact we can test this uniformity only for $0 < V(z) \leqslant V(z_{max}) = V_m$, say. In practice it is best to calculate V/V_m for each source in the sample and then look at the distribution of luminosity with respect to V/V_m. Fig. 6.12 shows such distributions for radio galaxies and for quasars in the third Cambridge catalogue of radio sources (see Table 5; p. 106). The distributions for quasars and for the more luminous radio galaxies are strikingly non-uniform (more sources are found at large values of V/V_m), showing that evolutionary effects are present. Similar results are obtained for other values of q_o.

6.9. INTEGRATED BACKGROUND RADIATION

The intensity of the integrated background radiation from a population of sources of luminosity P, and number density at the present epoch η_o, is

$$I = \int_{S=o}^{\infty} S \, dN(S) \; \mathrm{Wm}^{-2} \, \mathrm{sr}^{-1} \tag{6.28}$$

$$= \eta_o PR_o \int_{Z=o}^{\infty} \frac{dr}{z^2(1 - kr^2)^{1/2}} = \eta_o pc \int_{t=o}^{t_o} \frac{R(t)}{R_o} \, dt. \tag{6.29}$$

Using eqn (6.12):

$$I < \eta_o Pc\tau_o$$

for all models.

Fig. 6.13 shows the predicted background, using the Milne model, as a function of frequency for different populations of source, compared with the observed background.

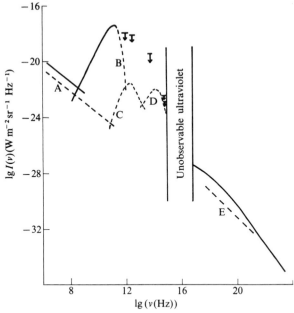

FIG. 6.13.   The intensity of the integrated background radiation.
Solid curves and upper limits: observations.   Broken curves:
predictions assuming no evolution.   A, radio galaxies and quasars;
B, 2·7 K cosmic black body; C, radiation from dust in galaxies;
D, integrated starlight in galaxies; E, rich clusters and
Seyferts.

Radio background

The main contribution is from the Milky Way, but it can be
shown that the steep-spectrum (mean spectral index $\alpha \sim 0\cdot 8$ -
section 1.4; p.16) extragalactic sources contribute an intensity
of about $10^{-21\cdot 6}$ W m^{-2} Hz^{-1} sr^{-1} at 178 MHz, about 10 times more
than that predicted assuming no evolution.   This confirms that
evolution has to be present, as was suggested by the source
counts.

Microwave and infrared background

The dominant observed background in the microwave region is
the 2·7 K black-body radiation.   In the submillimetre and

infrared region we have only upper limits at the moment. The
predicted contribution from interstellar dust in galaxies is
shown in Fig. 6.13.

Optical background

  The predicted integrated background from the starlight in
galaxies falls well below the observational limits. The problem
is that the brightness of the light from the earth's atmosphere
(for ground-based observations), from zodiacal light, and from
the starlight of the Milky Way together swamp out the cosmic
background. Possible changes with time in the luminosities of
galaxies due to stellar evolution lead to uncertainties in the
theoretical curves.

X-ray background

  There will be inverse Compton radiation (section 1.4; p.17)
from normal galaxies, from radio galaxies, and from quasars, due
to the interaction of the relativistic electrons responsible
for their radio synchrotron radiation with the photons of the
black-body radiation. However, unless the magnetic fields in
these objects are surprisingly low, the contribution to the
background is likely to be small. The main known contributors
to the background are the X-ray sources in rich clusters ($\sim$10 per
cent) and Seyfert galaxies ($\sim$20 per cent). The rest might be
inverse Compton radiation from relativistic electrons which have
leaked out from radio sources into intergalactic space, interacting with 2$\cdot$7 K black-body photons. Below 1 keV ($10^{17 \cdot 4}$Hz)
there may be a contribution due to hot intergalactic gas (see
Chapter 7). The observational status of the γ-ray background is
still controversial.

6.10. HORIZON

Even if we live in an open universe with infinitely many galaxies in it, the light from only a finite number of them will have reached us so far. There is therefore an *event horizon* which divides those events in the universe that we can know about, and those that we cannot yet know anything about. As time proceeds, galaxies swim into view. At first they are seen with very large red-shift. Since the expansion of the universe is slowing down ($q > 0$), the red-shift of any particular galaxy decreases with time.

PROBLEM

6.1. Work out the detailed predictions of the tests described in this chapter for the cases $q_0 = 0, 1/2$.

7. The density of matter in the universe

7.1. INTRODUCTION

We saw in Chapter 4 that whether the universe keeps on expanding indefinitely, or ultimately falls back together into a second fireball, depends on whether the average density of matter is less than or greater than the critical value

$$\rho_{ES} \sim 4 \times 10^{-27} \, (H_0/50)^2 \, \text{kg m}^{-3} \qquad (7.1)$$

(equivalent to $q_0 = 1/2$, the Einstein-de Sitter value - section 4.7). The cosmological tests described in the previous chapter have so far failed to settle this question, so we now try to determine the average density of matter in galaxies and in other possible forms directly. The most important of these other forms of matter is intergalactic gas, and this leads us naturally on to the question of the formation of galaxies.

To find the average density of matter we first need to determine the average mass of a galaxy and then multiply by the average number of galaxies per unit volume, determined by galaxy counts (section 6.7).

7.2. THE MASSES OF GALAXIES

A variety of methods have to be used, depending on the galaxy type.

(a) *Spirals*

Most of the material in the disc of spirals is moving in an approximately circular orbit in a balance between centrifugal force and gravity. Hence for material far out from the centre of the galaxy

$$V^2/r \sim GM/r^2 \quad \text{or} \quad M \sim rV^2/G. \qquad (7.2)$$

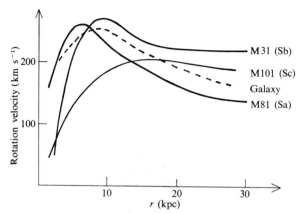

FIG. 7.1. The rotation velocity in different galaxies as a function of distance from the centre.

The rotation curve $V(r)$ can be determined by observing the Doppler-shifted 21-cm line of neutral hydrogen (Fig. 7.1).

(b) *Ellipticals*

Here we have to use what is known as the *virial theorem*, which tells us that for a system in equilibrium 2 × kinetic energy = gravitational energy, or $MV^2 \sim GM^2/r$, which is equivalent to eqn (7.2). We estimate the average kinetic energy per unit mass from the Doppler spreading of the emission lines in the spectrum of the nucleus of the galaxy.

(c) *Mass-to-light ratio*

By observing the ratio of mass to light for galaxies of different types, e.g. in binary systems, we can estimate the masses of individual galaxies from their total luminosity. For spirals $M/L \sim 7$, and for ellipticals $M/L \sim 50$, both in solar units. The problem here is that galaxies do not have sharply defined edges, so the total luminosity is hard to determine.

Methods (b) and (c) do not always give the same answer, in fact (b) usually leads to higher masses than (c). This could be due to 'missing mass' in the form of dwarf stars, black holes, or any other material that does not contribute significantly to the light of the galaxy

7.3. THE AVERAGE DENSITY OF MATTER IN GALAXIES

Fig. 7.2 shows the relative contribution to the average density of the universe of galaxies of different masses. Most of the mass comes in the form of high-mass galaxies, 10^{10}-$10^{12}M_{\odot}$. The curve has been extrapolated right down to isolated globular clusters, although it is not certain that the curve does not peak up again at this level.

The total density of matter in galaxies is then found to be

$$\rho_{g,0} \sim 3 \times 10^{-28}(H_0/50)^2 \text{ kg m}^{-3}, \qquad (7.3)$$

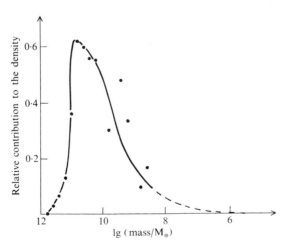

FIG. 7.2.   The relative contribution of galaxies of different masses to the average density of matter in the universe.

with an uncertainty of at least a factor 2. This is a factor
10 below the critical value (eqn (7.1)).

7.4. CLUSTERS OF GALAXIES

Methods (b) and (c) of section 7.2 can be used to estimate the
masses of clusters of galaxies. The virial theorem becomes

$$M = \overline{rv^2}/G, \qquad (7.4)$$

where the averages indicated by bars are over all the galaxies
in the cluster, and r now refers to the distance between the
galaxies. Typical masses are in the range 10^{13}-10^{15} $M_{\odot}$ and
typical mean densities inside clusters are 10^{-25} kg m^{-3}. The
latter can be presumed to be a firm upper limit on the average
density of matter in the universe.

Discrepancies between methods (b) and (c) of section 7.2, with
virial theorem masses up to a hundred times greater than visible
masses, again suggest 'missing matter', perhaps intergalactic
gas (which must be neither neutral nor too hot — see section
7.6). Both extended X-ray emission from rich clusters and the
appearance of radio 'tails' within clusters suggest that there
is intergalactic gas in clusters (section 2.9).

7.5. SOME OTHER POSSIBLE FORMS OF MATTER

The best candidate for matter with the critical density (eqn
(7.1)) is intergalactic gas, since we do not expect galaxy
formation to be a 100 per cent efficient process. This will be
discussed in the next section. Some other possible forms are
discussed below.

Dwarf galaxies

Although Fig. 7.2 suggests that the contribution from dwarf
galaxies to the average density of matter in the universe is

small, very few of these objects are known (in fact all are in
the Local Group) so their number density is very uncertain.
The estimate of Fig. 7.2 would have to be out by a factor of
more than 100 to arrive at the critical density. Whether
intergalactic globular clusters and isolated stars exist is still
unknown. A direct measurement of the optical background radiation
might help to decide this.

Compact objects and quasars

The simplest view about quasars and other compact objects (e.g.
N-galaxies) is that they represent outbursts in the nuclei of
galaxies. As such, their contribution to the average density of
matter would already have been included in eqn (7.3).

Even if they are a distinct class of object from galaxies,
quasars make a negligible contribution to the average density of
matter, unless a significant proportion of them are 'local'
objects with intrinsic red-shifts (see the Epilogue, p.144).

Dead galaxies and black holes

The normal types of galaxy surveyed in section 2.6 can be
expected to retain their present appearance for a time much longer
than the Hubble time. However, there is the possibility of an
earlier generation of galaxies now no longer visible due to
exhaustion of stellar nuclear energy sources (they would have to
have formed only massive stars). Another possibility that cannot
be ruled out is that there are massive black holes spread
throughout the universe. Either could easily contribute the
critical density.

Planets, rocks, and dust

Only weak limits can be placed on the amount of cool, solid
matter in the universe. If the average density of the grains
$\rho_d \sim 10^3$ kg m^{-3}, and their contribution to the average density of

the universe is given by eqn (7.1), then in order to be able to see a Hubble distance without drastic absorption, the grains need to be larger than 1 mm.

$$(\frac{4\pi}{3}a^3 \rho_d \eta_d \sim 4 \times 10^{-27}, \eta_d \pi a^2 c\tau_0 < 1,$$

where a, η_d are the radius and number density of the grains.) Thus there is no real observational limit on the amount of matter in the universe in the form of large dust, rocks, or planets.

Neutrinos

We mentioned briefly in section 5.3 that cosmic neutrinos and antineutrinos are expected to become decoupled from matter and radiation at an early stage in the fireball. The expected energy density is about 20 per cent of that of photons, i.e. negligible compared to matter at the present epoch. This neutrino flux is completely undetectable with present techniques (section 1.3).

Gravitational radiation

It would be possible for this to be the dominant form of energy at the present epoch: current gravitational wave detectors are too insensitive by a large factor to test this.

Cosmic rays

The local energy density of cosmic rays ρ_{cr} is about the same as that of the microwave background and corresponds to

$$\rho_{cr} \sim 1 \cdot 6 \times 10^{-30} kg \ m^{-3} \qquad (7.4)$$

$$\ll \rho_{ES}.$$

Even this value may hold only within our Galaxy. In fact cosmic-ray nuclei with energies $< 10^{17}$ eV are confined to our Galaxy by its magnetic field, so are almost certainly of Galactic origin. And the average energy density of cosmic-ray electrons must be less than a thousandth of the local value, otherwise inverse Compton interaction with the microwave background and integrated starlight photons would give too large an X-ray background (section 1.4, p.17). However, the higher-energy cosmic-ray nuclei leak out of our Galaxy in only about 3×10^6 years (this can be deduced from the abundance of the cosmic-ray-created elements Li, Be, B), so the density (eqn (7.4)) may well be universal. If it is, then we cannot have intergalactic gas at the critical density (eqn (7.1)), since (i) the cosmic rays would heat it up so much that too many X-rays would be produced, (ii) the production of γ-rays from cosmic ray secondary π^0 meson decay would exceed the observed limit in the range 50-100 MeV.

7.6. INTERGALACTIC GAS

If there is intergalactic gas with the critical density (eqn (7.1)), then rather stringent limits can be set on its physical state.

Atomic hydrogen (H_I)

This would be observable by the following.

(a) *21-cm absorption*: failure to see this in the radio-galaxy Cyg A shows that the density of intergalactic neutral hydrogen

$$\rho_{H_I} < 10^{-27} \text{ kg m}^{-3}.$$

(b) *21-cm emission*: again, this has not been observed.

(c) *Soft X-ray absorption*: an X-ray photon ionizes a hydrogen (or helium) atom, which then recombines, but emits a less energetic photon. The opacity of hydrogen for $\lambda < 912$ Å is proportional to $(\lambda/912)^3$, and failure to see absorption down to 50 $\overset{o}{A}$ shows that

$$\rho_{H_I} < 10^{-28} \text{ kg m}^{-3}.$$

(d) *Lyman-α absorption*: a trough should appear on the long-wavelength side of the Lyman-α line at 912 $\overset{o}{A}$. For quasars this is conveniently red-shifted into the visible range, and the absence of such a trough shows that

$$\rho_{H_I} < 10^{-33} \text{ kg m}^{-3}$$

if quasar red-shifts are cosmological. Thus if hydrogen is present with density as defined in eqn (7.1), it must be highly ionized.

Ionized hydrogen (H_{II})

We are likely to have $T > 10^5$ K since a cold, ionized gas recombines very quickly, and ionizing mechanisms, e.g. ultraviolet photons or cosmic rays produced by galaxies and quasars, tend to heat the gas too. The re-heating of the gas cannot have occurred too soon after decoupling (section 5.2), since observable distortions in the 2·7 K black-body spectrum would have been produced.

In a hot ionized gas free electrons moving under the influence of each others' electrostatic fields radiate *free-free* (or *thermal Bremsstrahlung*) radiation. The observed radio and X-ray backgrounds give strong limits on the temperature:

Radio (20 cm): the gas can only have been heated to 10^5 - 10^6 K at red-shifts $z < 100$.

Hard X-rays $(E > 1 \text{ keV})$: $T < 3 \times 10^8$ K if $\rho = \rho_{ES}$.

Soft X-rays $(E = 0 \cdot 25 \text{ keV})$: there is a possible detection of gas with

$$\rho \sim \rho_{ES}, \quad T \sim 10^6 \text{ K}.$$

If this hot intergalactic gas exists, then it attenuates the light from distant sources (but not the integrated background) by Thompson (i.e. free-electron) scattering, by a factor $\exp(-\tau_e)$, where

$$\tau_e = 0 \cdot 023 \left\{ (1 + z)^{3/2} - 1 \right\}$$

in the Einstein-de Sitter model.

Other weak effects are the absorption of low-frequency radio waves, (Faraday) rotation of the plane of polarization of distant sources if there is also an intergalactic magnetic field, and a frequency-dependent time-lag in the arrival of light from variable sources (*dispersion*).

Molecular hydrogen

This cannot make up a very significant fraction of the matter in the universe, due to the absence of a Lyman-α trough in quasar spectra, although it has recently been recognized that molecular hydrogen makes up a major part of the gas in our Galaxy.

7.7. FORMATION OF GALAXIES

At the epoch of recombination (section 5.2), density fluctuations in the matter distribution have the chance to form galaxies by gravitational contraction. They can do so provided the time-scale for gravitational contraction is less than the time for sound waves to cross the irregularity, i.e. if the size of the fluctuation

$$L > v_s (G\rho)^{-1/2} = L_J, \tag{7.5}$$

where L_J is the *Jeans length*, v_s is the velocity of sound, and ρ is the density of the gas, since it can be shown that the time for gravitational contraction is $(G\rho)^{-1/2}$.

The corresponding critical (or *Jeans*) mass turns out to be about $10^5 \ M_{\odot}$, so that unless higher mass irregularities are already fairly well developed ($\Delta\rho/\rho > 1$ per cent), the whole gas would tend to break up into globular-cluster-sized objects rather than galaxies. This is a serious difficulty for big-bang cosmology: to form galaxies you have to postulate substantial irregularities there to start with. An alternative approach will be mentioned in section 8.5.

PROBLEMS

7.1. Test how well the galaxies of the Local Group (Table 1; p.4) agree with Fig. 7.2. What reasons can you think of for less than perfect agreement?

7.2. Apply the virial theorem (eqn (7.4)) to the Local Group of galaxies (assume the total velocity of the galaxies is $\sqrt{3}$ times the radial velocity). How well does the virial-theorem mass agree with the total observed mass, and why?

8. Other cosmological theories

8.1. GENERAL RELATIVISTIC MODELS WITH THE Λ-TERM

When Einstein originally put forward his general theory of
relativity, he included an additional term in the field equations,
the so-called *cosmological term*. This modifies the law of
gravitation at large distances into an attraction or repulsion
directly proportional to distance, $\ddot{\underline{r}} = \Lambda\, \underline{r}$, Λ constant. No such
effect is observed in the solar system, or in the structure of our
Galaxy, so Λ must be very small. This extra term would have an
effect only on the scale of clusters of galaxies or larger, hence
the name for Λ of *cosmological constant*.

The Λ-term is consistent with all the basic principles that led
Einstein to his field equations (it is effectively a constant of
integration), but it is usually set equal to zero by relativists
in order to keep the theory as simple as possible. However, it
leads to some new cosmological possibilities, which we will now
investigate.

The equations for the scale factor $R(t)$, (4.11) and (4.12)
(p.70), which are derived from the field equations assuming the
cosmological principle, become

$$\ddot{R} = -4\pi G\rho_0 R_0^3/3R^2 + \Lambda R/3, \qquad (8.1)$$

$$\dot{R}^2 = 8\pi G\rho_0 R_0^3/3R - kc^2 + \Lambda R^2/3 = G(R). \quad (8.2)$$

First we see that the Λ-term does not have any effect near $R = 0$,
so behaviour near the 'big bang' is unaltered. We consider the
cases $\Lambda < 0$ and $\Lambda > 0$ separately:

$\Lambda < 0$

R has to be finite for $\dot{R}$ to remain a real number, and there

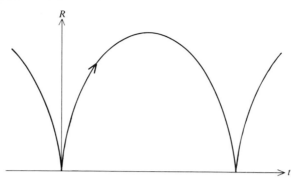

FIG. 8.1. $\Lambda < 0$ models.

exists an R_c such that $G(R_c) = 0$, i.e. $\dot{R} = 0$ when $R = R_c$. Eqn (8.1) then shows that $\ddot{R} < 0$ when $R = R_c$, so the universe starts to contract at this point. We therefore have oscillating models if $\Lambda < 0$, whatever the value of k (see Fig. 8.1).

$\Lambda > 0$

If $k \leqslant 0$, $\dot{R}^2 > 0$ all R, so we have a monotonic expanding universe, the only difference from those with $\Lambda = 0$ being that at large R, $\dot{R}^2 \sim \Lambda R^2/3$, so

$$R \propto \exp \sqrt{(\Lambda/3)}\,t. \qquad (8.3)$$

If $k = 0$, $\rho_o = 0$, $\Lambda > 0$ we have the de Sitter model, for which eqn (8.3) holds for all t.

$k = 1$

In this case there is a critical value of Λ, Λ_c, such that $\dot{R} = 0$ and $\ddot{R} = 0$ can both be satisfied simultaneously. From eqn (8.1), $\ddot{R} = 0$ implies

$$R = R_o (4\pi G\rho_o/\Lambda)^{1/3} = R_c, \text{ say}, \qquad (8.4)$$

and then eqn (8.2) implies that

$$0 = (4\pi G\rho_o)^{2/3} \Lambda^{1/3} R_o^2 - kc^2$$

so,

$$\Lambda_c = (kc^2)^3/R_o^6 (4\pi G\rho_o)^2. \qquad (8.5)$$

This means that there is the possibility of a static model of the universe, with $R = R_c$, $\Lambda = \Lambda_c$, for all time t, provided

$$\Lambda = 4\pi G\rho_c = kc^2/R_c^2, \qquad (8.6)$$

and since $\rho_c > 0$, k must be positive for this to happen. This is the Einstein static model, the first solution of general relativity to be found that satisfies the cosmological principle.

By studying the function $G(R)$ (eqn (8.2)) as a function of R, we can see what other possible $\Lambda > 0$, $k = +1$, models there are (Fig. 8.2). Clearly $G(R) \to \infty$ both for $R \to 0$ and for $R \to \infty$, and reaches a minimum at R_c, with $G(R_c) >$ or < 0 according as $\Lambda >$ or $< \Lambda_c$.

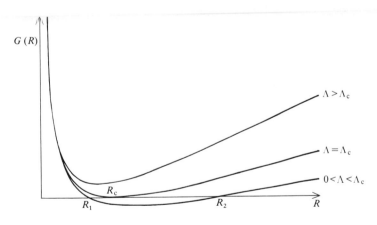

FIG. 8.2. $G(R)$ for $\Lambda > 0$, $k = +1$ models.

$\Lambda > \Lambda_c$

$G(R) > 0$ all R, so we have a monotonic expanding universe again.

$\Lambda = \Lambda_c$

Apart from the Einstein static model, there are two models that approach this asymptotically, corresponding to the two branches of $G(R)$ — see Fig. 8.2. One expands out gradually from the Einstein state at $t = -\infty$ and then turns into an exponential expansion (eqn (8.3)). The other expands out from the usual big bang and then tends asymptotically to the Einstein model as $t \to \infty$. These are called the Eddington-Lemaître models (Fig. 8.3), EL1 and EL2.

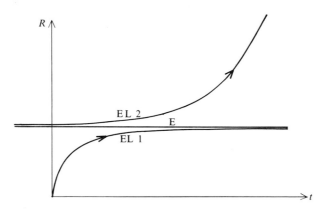

FIG. 8.3. $k = +1$, $\Lambda = \Lambda_c$ models. E = Einstein static model; EL1, EL2 = Eddington-Lemaître models.

If $\Lambda = \Lambda_c (1 + \varepsilon)$, $\varepsilon \ll 1$, we have the Lemaître models. For a long period of time R is close to R_c and the cosmological repulsion and gravitational attraction are almost in balance. Finally the repulsion wins and the expansion continues again.

$0 < \Lambda < \Lambda_c$

There are no solutions for $R_1 < R < R_2$ (Fig. 8.2). The solution with $R \lesssim R_1$ is an oscillating model. In the one with $R \gtrsim R_2$, the universe 'bounces' under the action of the cosmological repulsion (Fig. 8.4).

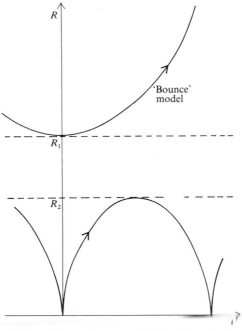

FIG. 8.4. $k = +1$, $0 < \Lambda < \Lambda_c$ models.

3.2. OBSERVABLE CONSEQUENCES OF THE Λ-TERM:

The Einstein static model can be eliminated immediately since it does not predict a red-shift. There remain two models which do not originate in a big bang, the EL2 (Fig. 8.3) and 'bounce' models. In each case there would be a maximum red-shift defined by

$$1 + z_{max} = R_0/R_{min},$$

and in the bounce models more distant objects would show a blue-shift. If quasar red-shifts are cosmological then $z_{max} \gtrsim 4$.

The Lemaître models permit ages of the universe far greater than the Hubble time τ_0 (the EL2, bounce, and de Sitter models have infinite ages). Since these models have positive curvature

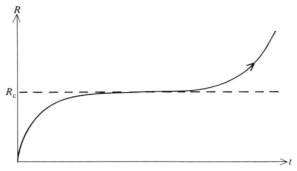

FIG. 8.5.   Lemaître model, with long 'quasi—stationary' period.

(see Fig. 4.5, p.75) and are spatially finite, there is the
intriguing possibility of seeing all the way round the universe,
and even of seeing a ghost Milky Way (normally light does not
have time to make this circumnavigation, resulting in a
horizon — section 6.10). So far we have no evidence that this
is happening. Another effect of the long 'coasting' period
(Fig. 8.5) is that there would be a concentration of objects
with red-shifts given by

$$1 + z \sim R_0/R_c,$$

and this has been invoked to explain the concentration of quasar
red-shifts at $z \sim 2$.

The best evidence for non-zero Λ would be if the parameters
σ and q were unequal at the present epoch, since eqn (8.1)
implies that

$$\Lambda/3 = H_0^2(\sigma_0 - q_0). \qquad (8.7)$$

We have seen that the observed matter in galaxies corresponds to
$\sigma_0 \sim 0.1$ (eqn (7.3)), whereas the magnitude red-shift for
bright galaxies gave $q_0 \sim 1$, but both these estimates are far
too uncertain to be interpreted as implying $\Lambda \neq 0$.

The condition $\Lambda = \Lambda_c$, which defines the Eddington-Lemaître

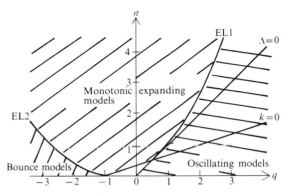

FIG. 8.6. The regions of the σ-q plane occupied by different types of cosmological model.

models, can be shown to be equivalent to

$$(3\sigma - q - 1)^3 = 27\sigma^2(\sigma - q). \tag{8.8}$$

The zones of the different models in the $\sigma - q$ diagram are shown in Fig. 8.6.

8.3. THE STEADY STATE COSMOLOGY

This was put forward in 1948 by Bondi, Gold, and Hoyle. The cosmological principle was strengthened to the 'perfect' cosmological principle: the universe presents the same appearance to all fundamental observers at all times.

An immediate consequence of this is that the Hubble time must be a constant:

$$\tau = R/\dot{R}, \text{ constant for all t.}$$

Thus

$$R \propto \exp t/\tau, \text{ or } R = R_0 \exp (t-t_0)/\tau. \tag{8.9}$$

A second consequence is that the density of matter is constant, and to maintain this we must have continuous creation of matter at a steady rate per unit volume

$$\frac{1}{R^3} \frac{d}{dt} (\rho R^3) = 3 \rho/\tau. \tag{8.10}$$

It can also be shown that the three-spaces t = constant have a three-dimensional curvature, known as the Guassian curvature, of $k/R(t)^2$, which would depend on time unless k = 0.

We have therefore shown that the steady-state metric is

$$ds^2 = dt^2 - \frac{\exp(2t/\tau)}{c^2}(dx^2 + dy^2 + dz^2), \qquad (8.11)$$

and this is the same as that for the de Sitter model (section 8.1).

If field equations similar to those of general relativity are used, but with an extra term representing the creation of matter, it is found that

$$8\pi G\rho\tau^2/3 = 1, \qquad (8.12)$$

which with $\tau \sim 2 \times 10^{10}$ years gives $\rho \sim 4 \times 10^{-27}$ kg m^{-3}, about 10 times higher than the density of matter in galaxies (eqn (7.3)) The remainder could be made up of ionized intergalactic hydrogen at a temperature $\sim 10^6$K (section 7.6). The corresponding creation rate (of, presumably, cold, neutral, uniform hydrogen) is the undetectable 10^{-44} kg m^{-3}s^{-1}. The magnitude-red-shift relation for the steady-state theory is the same as for the de Sitter model:

$$m = A + 5\lg\{z(1 + z)\} \qquad (8.13)$$

(where A is constant), equivalent to q_0 = -1, which is rather improbable from the data on bright cluster galaxies (Fig. 6.4, p.100), bearing in mind that no evolution is permitted.

The number of sources per steradian out to red-shift z is

$$N(z) = \int_0^{r_0(z)} \eta R^3 r^2 \, dr = \eta(c\tau)^3 \int_0^z (1 + z)^{-3} z^2 dz$$
$$= \eta(c\tau)^3 \left\{ \ln (1 + z) - \frac{2z + 3z^2}{2(1 + z)^2} \right\}. \qquad (8.14)$$

The corresponding $N(S)$ curve is flatter than -1·5, whereas the radio-source counts are appreciably steeper (Fig. 6.11, p.107).

The final blow for the steady-state theory was the isotropic 2·7 K black-body radiation, for which no convincing explanation was produced, whereas for big-bang cosmology this is one of its greatest successes.

However, it should be remembered that in 1948, the accepted estimate of the Hubble time was only 2×10^9 years, only 10 per cent of the age of our Galaxy. Steady-state cosmology provided an elegant way out of that difficulty, now resolved by new measurements of τ_o.

8.4. THEORIES IN WHICH G CHANGES WITH TIME

Theories of this type were first proposed by Dirac and Jordan. More recently Brans and Dicke, Hoyle and Narlikar, and Dirac have put forward more elaborate theories of this type.

A variation of G with time has a considerable effect on the evolution of the earth and solar system. It has been suggested that the continents all fitted together at one time on a much smaller earth. As the gravitational constant reduced, the earth expanded to its present size and the continents were forced apart. Also a star in its hydrogen-burning phase, like the sun, has a luminosity

$$L \propto G^7, \tag{8.15}$$

and so would have been appreciably brighter in the past. The effect of this on life on earth would be enhanced by the fact that the earth must be moving away from the sun if G is declining. It has been argued that G cannot vary with time faster than $t^{-1/4}$ without causing the sun to have already evolved to a red giant.

Similar changes would occur for our Galaxy as a whole; it would have been smaller and more luminous in the past. The effect on the magnitude-red-shift relation would be appreciable. Sandage's linear relation (Fig. 6.4; p.100) would be a most remarkable coincidence, a combination of the luminosity evolution due to the variation of G and geometrical effects exactly cancelling out.

The best upper limit to date from solar-system studies on the variation of G is

$$\dot{G}/G < (2.5 \times 10^9 \text{ years})^{-1}, \qquad (8.16)$$

not yet sufficient to rule out this type of theory.

The Brans-Dicke cosmology represents the simplest extension of general relativity. In addition to the tensor gravitational field represented by the metric tensor (section 4.4), there is a scalar field (the *gravitational 'constant'*) which is a function of time only.

Brans and Dicke take $\Lambda = 0$ and seek to satisfy *Mach's principle*, that local inertial properties should be determined by the gravitational field of the rest of the matter in the Universe, by taking

$$G^{-1} \sim \sum_{\text{universe}} m/rc^2. \qquad (8.17)$$

The models for the case $k = 0$ are particularly simple, since

$$R \propto t^q, \quad G \propto t^{-r},$$

where

$$q = 2(1 + \omega)/(4 + 3\omega), \quad r = 2/(4 + 3\omega) \qquad (8.18)$$

and ω is a 'coupling constant' between the scalar field and the geometry. $\omega \to \infty$ gives the Einstein-de Sitter model. Note that for general ω, $G\rho t^2 = $ constant. Dirac's 1937 theory is obtained by setting $\omega = -2/3$.

The Brans-Dicke theory predicts slightly different amounts for the deflection of light by the sun and for the perihelion advance of a planet, and these should be conclusively tested soon. So far we can say $\omega \gtrsim 6$, which means the models differ negligibly from those of general relativity in the behaviour of $R(t)$. The variation in the luminosity of a hydrogen-burning star, and hence of a galaxy, would be

$$L \propto t^{-n}, \text{ with } n \leqslant 7/11, \tag{8.19}$$

and this is no greater than the amount already expected due to the gradual exhaustion of gas in galaxies available for forming new stars.

In the Hoyle and Narlikar and Dirac (1973) theories the solar system experiments would give the same results as general relativity. The variation of G arises because there are two time-scales, atomic time and cosmological time, which no longer coincide.

8.5. ANISOTROPIC UNIVERSES IN GENERAL RELATIVITY

The high degree of isotropy of the 2·7 K black-body radiation made it natural to consider isotropic universes. However, it is possible to have a universe which is highly anisotropic early on but which would look almost exactly isotropic now. The initial singularity might, instead of being point-like, be like a 'cigar' or a 'pancake', or the universe might oscillate between these for a while.

Misner has suggested that the universe started off completely anisotropic and inhomogeneous. The anisotropy resulted in the universe being thoroughly mixed, and interaction between neutrinos acted like a viscosity to smooth out most of the inhomogeneities, leaving precisely those that we want, namely, galaxies and clusters. Another way of damping out anisotropy that has been put forward by Zeldovich and other Russian cosmologists is through particle production from the tidal energy associated with the anisotropy. These ideas are attractive, not only for providing an explanation of how galaxies form (one of the weaknesses of big-bang cosmology), but also for giving some explanation of the cosmological principle. The mathematical difficulties of developing these ideas are formidable.

8.6. THE HIERARCHICAL UNIVERSE

Another example of a model with a totally different philosophy to the big-bang models is the hierarchical universe advocated by de Vaucouleurs and others.

The idea is that the hierarchy of condensation that proceeds: stars - clusters of stars - galaxies - clusters of galaxies, simply carries on *ad infinitumm* with average density in a condensation decreasing as the scale of the condensation increases, and ultimately tending to zero.

The question whether clusters of clusters exist is still a controversial one, especially as one astronomer's 'cluster of clusters' is another's 'large cluster with sub-condensations'. There are certainly complexes containing two or more clusters. But if all galaxies and clusters in the observable universe belong to a 'metacluster', then any other similar ones must be exceedingly remote, otherwise we would see inhomogeneities in the black-body radiation.

8.7. EDDINGTON'S MAGIC NUMBERS

Eddington remarked that if you work out the ratios

$$\frac{\text{electromagnetic force between proton and electron}}{\text{gravitational force between proton and electron}} =$$

$$\frac{e^2}{G\,m_e m_p} = 0 \cdot 23 \times 10^{40}$$

and

$$\frac{\text{radius of universe}}{\text{classical electron radius}} = \frac{c\tau_0}{e^2/m_e c^2} \sim 8 \times 10^{40},$$

you have two huge dimensionless numbers of the same order of magnitude and that it would be surprising if this were a coincidence. If $G\rho_0\tau_0{}^2 \sim 1$, then a third dimensionless number can be deduced:

$$\left\{ \frac{\rho_0(c\tau_0)^3}{m_p} \right\}^{\frac{1}{2}} \sim 10^{40},$$

which is roughly the square-root of the number of particles in the universe, if $k = 1$.

Two interpretations are possible. (i) We are at a special epoch, at which these numbers happen to be the same. It has been suggested, for example, that we *are* at a special epoch, namely, the one at which life has evolved to the state that these questions can be asked. Certain conditions are necessary to produce life and these involve relationships between the fundamental constants (this has been called the 'anthropic principle'). (2) These numbers remain equal at all epochs, so that some of the fundamental constants change with time. In Dirac's 1937 theory

$$G \propto t^{-1}, \quad R \propto t^{1/3}, \quad \tau \propto t, \quad \text{and} \quad G\rho t^2 = \text{constant},$$

so all three dimensionless numbers are proportional to t and remain in the same proportions at all times. However, this leads to too short an age for the universe. Gamov has suggested $e^2 \propto t^{-1}$ (G = constant), as an alternative, but this runs into difficulties with isotope abundances and with the fact that the lines in quasar and galaxy spectra show that the *fine-structure constant* e^2/hc is not varying with epoch.

PROBLEM

8.1. Work out the diameter-red-shift relation and the integrated background radiation in (a) the de Sitter model (b) the steady-state model.

Epilogue: twenty controversies in cosmology

In this epilogue are outlined some of the cosmological
controversies and issues not yet totally resolved. It is a
personal perspective, as of June 1975. Other cosmologists might
consider many of these questions settled, or suggest other items
for inclusion. But I believe it to be important to emphasize
that science is more to do with argument and controversy than
a sacred canon of 'knowledge' handed on from generation to
generation.

1. GENERAL RELATIVITY

Most of the models of the universe described in this book are
based on general relativity, which cannot be said to rest on a
very solid experimental basis. The solar system tests of general
relativity have been successful so far, to an accuracy of better
than 10 per cent, but these test the theory only in a weak
gravitational field. It is important to increase the accuracy of
these tests, and thus rule out the Brans-Dicke theory, which
predicts slightly different results to general relativity. Even
more important is the possibility of finding black holes in
binary systems through X-ray emission from hot in-falling gas,
although there must always remain some doubts about the interpretation of these systems. The best candidate to date is the X-ray
source Cyg X-1. We know that general relativity must break down
when densities become so high that quantum gravitational effects
become important. Thus we do not yet have an acceptable theory
of the earliest stages of the big bang, and the study of quantum
gravity and of singularities remain outstanding problems in
cosmology.

2. VARIATION OF G

Perhaps the most likely way that general relativity might be proved wrong is through variation in the gravitational constant G with time. This can be studied both by variation in the moon-earth distance and by possible changes in the radius of the earth with time. Models which predict that G varies with time were discussed in section 8.4.

3. ISOTROPY OF THE UNIVERSE

The limits on anisotropy of the microwave background radiation, 0·1 per cent at the moment, place severe limits on the anisotropy of the universe. It remains a matter of controversy *why* the universe should be so isotropic. Within the framework of general relativity, a first answer appears to be that it is simply a consequence of the initial conditions in the universe. Mach's principle, to which much importance has been attached by supporters of the steady-state theory, and which is a major part of the motivation for the Brans-Dicke theory, was shown by Gödel not to be incorporated into general relativity. This principle relates the local inertial frame to the large-scale distribution of matter in the universe and would forbid the arbitrary relative rotation between these found by Gödel to be permissible in general relativity. Anisotropic expansion, however, would still be permitted.

Misner has suggested that the universe started off highly anisotropic and inhomogeneous but evolved to isotropy and homogeneity on the large scale (section 8.6). Hawking, on the other hand, argues that if the universe were not isotropic we would not be here, since galaxies would not be able to form in an anisotropic universe. Thus the universe is isotropic *because we are here.*

McCrea, in a novel extension of the Machian viewpoint, argues that since information reaching us now from some remote part of the universe set off far in the past, and is degraded by the red-shift, a *principle of uncertainty* has to be introduced. Our knowledge of the universe (and hence whether we can say it is homogeneous and isotropic) deteriorates as we extrapolate to great distances.

4. THE AGE OF THE UNIVERSE

The age of the oldest stars in our Galaxy is about $1 \cdot 5 \times 10^{10}$ years, with an uncertainty of ± 30 per cent. If there was a big bang it must have happened longer ago than that and, if $\Lambda = 0$, this implies $H_0 < 60$ km s^{-1} Mpc^{-1}. In Fig. 3.1. (p.51) we saw that H_0 almost certainly lies between 25 and 100, with a best value of around 50. This gives a rather narrow margin: our Galaxy must have formed early on in the history of the universe (at an epoch corresponding to $z > 5$ if $q_0 \ll 1$). If $q_0 = 1/2$, then the age of the universe has to be $< 1 \cdot 2 \times 10^{10}$ years.

Clearly it is important to get better information on the Hubble constant, the age of the Galaxy, and the epoch of galaxy formation. The possibility of a discrepancy similar to the one that led to the invention of the steady state theory still can not be totally ruled out.

5. THE VELOCITY-DISTANCE RELATION

De Vaucouleurs claims that this relation is not linear but quadratic ($v \propto d^2$) at small distances (out to the Virgo cluster). He attributes this to the effect of the local disc-like concentration of matter which he calls the Local Supercluster, centred on Virgo with a radius of about 25 Mpc. Our Galaxy would therefore be on the edge of it.

De Vaucouleurs obtained this result using the diameters of small groups of galaxies as a distance indicator. Sandage has argued vigorously both against this non-linearity and against the hierarchical models favoured by de Vaucouleurs. The existence of the Local Supercluster is more widely accepted, on the other hand, although it poses the problem of why we should be in so unusual a system. The distance scale would have to be underestimated by a factor of 10 for other rich clusters to have the same dimensions, i.e. $H_o \sim 5$! Alternatively, distant clusters must be much larger objects than they appear (this has been argued by Peebles).

Again we see how important it is to improve the distance scale, especially at distances beyond the Virgo cluster. Observations of supernovae may be the best way of doing this.

6. CLUSTERS OF CLUSTERS

A controversy related to the previous one is whether clusters of galaxies are themselves clustered. Conglomerations containing several rich clusters have certainly been found, although it is not clear how common they are, or whether they are really single clusters with several sub-condensations.

The main points of view taken on this subject are: (a) There are essentially no clusters of clusters, and the universe is smooth on a scale larger than the average distance between clusters (Zwicky); (b) there really do exist clusters of clusters (Abell, de Vaucouleurs): (c) clustering occurs on all scales so there is no limit to how large a cluster can be (Kiang).

The possibility that most of the galaxies we can see form part of some gigantic Metacluster cannot be ruled out at the moment, although limits can be set on how frequent such Metaclusters are in the universe (from the radio background) and on the state they were in at the epoch of recombination (from the absence of fluctuations in the microwave background).

7. ANTIMATTER

In our vicinity there is a strong preponderance of matter over antimatter, and it is normally assumed that this is true for the universe as a whole, as a result of initial conditions. Why should this be so? Some cosmologists, especially Alfven and Omnes, have argued that there are equal amounts of matter and antimatter in the universe, but that they remain spatially segregated. The encounter of lumps of matter and antimatter (whole galaxies perhaps) would yield enormous amounts of annihilation energy.

While there is little evidence for this point of view, one prediction which should be tested soon is that annihilation occuring during the decoupling era should lead to a strong γ-ray background excess above 70 MeV, due to decay of π^0 mesons.

8. PRIMORDIAL HELIUM

Most of the oldest stars in our Galaxy contain about 27 per cent helium by mass, normally assumed to be the result of nuclear reactions during the fireball. However, there are some old stars with very weak or non-existent helium lines in their spectra. This could either be because helium has sunk out of their atmospheres, or because there is no primordial helium.

In the latter case, an early generation of massive stars must have existed in our Galaxy in which the helium found in the later generations was formed. The history of the fireball would also have to be modified so as to predict no primordial helium, e.g. by anisotropy.

9. THE MOTION OF THE EARTH WITH RESPECT TO THE MICROWAVE BACKGROUND

The net motion of the earth with respect to the local frame in which the universe is isotropic, due to our motion round the sun,

the orbit of the sun round the Galaxy, and of our Galaxy through
the Local Group, should soon be detected through a dipole
anisotropy in the black-body radiation. The sky will appear
hotter in the direction we are going towards and colder in the
direction we are coming from.

Until this is found we have to remain slightly anxious that
this is not another Michelson-Morley experiment. If no anisotropy
greater than 0·01 per cent is found, there must be some flaw in
our application of special and general relativity to the
cosmological scale.

10. IS THE MICROWAVE BACKGROUND A RELIC OF THE FIREBALL?

This background, with a 2·7 K black-body spectrum over the
range of wavelengths 1 mm $< \lambda <$ 30 cm, is the main evidence for
a phase of the universe in which radiation was the dominant form
of energy, the fireball.

It is still just about possible, however, to explain the
background in terms of the superposition of discrete sources spread
throughout the universe. The shape of the spectrum and the great
smoothness of the background require a very strong and special
evolution in the properties of this population of discrete sources,
which would radiate mainly in the submillimetre region. Improved
information on the spectrum and smoothness of the background may
rule out such models.

11. FORMATION OF GALAXIES

This is a very poorly understood topic, and it is not even
known whether our Galaxy started to look something like its
present shape at epochs corresponding to red-shifts 1, 10, or 100.
The smoothness of the black-body radiation tells us that
condensations had hardly started at all at the epoch of decoupling
$(z \sim 1000)$.

12. DID ALL GALAXIES FORM AT THE SAME TIME?

Although the properties of the main type of galaxy in the Hubble
sequence can be understood in terms of objects all of the same age
forming stars at different rates, there is not much real evidence
for this. We can see star clusters in some very nearby
galaxies and the turn-off point on the main sequence in these
suggests similar ages to our own Galaxy, but it is hardly
surprising to learn that Local Group galaxies formed at the same
time.

If star formation in ellipticals proceeded at the same rate
as in our Galaxy, they would have to be about 10 times older than
our Galaxy to have so little gas left. This would be inconsistent
with our basic big-bang scenario.

13. IS THERE AN INTERGALACTIC MEDIUM?

This was discussed in section 7.6, and we saw that there could
be the critical density of intergalactic hydrogen, provided it is
at a temperature of about a million degrees. Soft X-ray emission
provides the best hope of detecting such gas, although it is hard
to decide how much emission is from gas in our own Galaxy.

The problems of heating such gas, if it exists, and of determining the epoch at which it was heated, remain unsolved, although
some constraint can be set by radio, microwave, and X-ray
background observations.

14. MISSING MASS

The question whether clusters of galaxies are bound together
by the self-gravitation of invisible matter remain a controversial
one. If they are not bound, are they a transient phenomenon, or
did some violent events cause them to start to break apart?

15. RADIO-SOURCE COUNTS

Most workers believe that the low-frequency radio-source counts imply strong evolutionary effects in the radio-source population (mainly radio galaxies, with about 25 per cent quasars) and that the high-frequency ($\nu > 1000$ MHz) source counts can be understood if suitable assumptions are made about the flat spectrum ($\alpha < 0\cdot5$) sources which play a more important role at high frequencies.

However, some atronomers, notably Hoyle and Kellerman, argue that (a) the low-frequency counts are steep because of some local fluctuation, (b) the high-frequency counts do not imply evolution. This can probably be settled only when high-frequency surveys have been studied in the same detail as low-frequency ones.

16. COUNTS OF GALAXIES

The evidence for a homogeneous distribution of galaxies throughout the universe rests unsteadily on counts of galaxies, which are laborious to perform and hard to interpret. It appears that present available data can be interpreted as supporting both a uniform distribution of galaxies *and* very strong evolutionary effects (inconsistent with our basic big-bang scenario). Counts of clusters of galaxies lead to the same ambiguity.

17. THE NATURE OF THE X- AND γ-RAY BACKGROUND

Since this is the only background which may come wholly from discrete sources, it should provide important information about the distribution of matter at large red-shifts and evolutionary effects. However, the number of theories at present held is almost as great as the number of workers in the field. Detailed spectra of individual sources will provide important constraints and the next generation of X-ray telescopes will greatly advance the field.

18. THE DISTANCE OF QUASARS, NON-COSMOLOGICAL RED-SHIFTS

The three main points of view on quasars are (a) they are simply galaxies in which a violent outburst has occurred and their red-shifts are cosmological, (b) there is a subset of quasars with intrinsic red-shifts, which are therefore much weaker and nearer objects (c) all quasars and many galaxies have intrinsic red-shift components.

In favour of (a) is the fact that some quasars are in groups of galaxies with the same red-shift as the quasar, that some quasars have faint nebulosity surrounding the stellar image of about the right size to be a parent galaxy, and that there is a correlation of apparent radio size with red-shift (Fig. 6.8). As evidence for (c), Arp, the Burbidges, and co-workers have found many examples of objects close together on the sky with very different red-shifts. Examples include peculiar galaxies associated with radio sources, chains of galaxies and of quasars, quasars associated with nearby spiral galaxies, and small companion galaxies near larger ones. Supporters of (a) argue that every one of these associations is purely due to chance super-positions of objects at different distances. Under hypothesis (b), the arguments that support (a) are accepted, but it is suggested that the subclass of quasars with compact, variable radio sources are in fact weak objects similar to Seyfert nuclei and therefore that their red-shifts are mainly intrinsic, possibly gravitational. The gravitational red-shift could not explain the intrinsic red-shifts required in extended objects under (c).

19. COINCIDENCES IN ENERGY DENSITIES

It was pointed out by Hoyle that the energy densities of the microwave background, of cosmic rays, of the magnetic field in our Galaxy, and of starlight in our Galaxy are all of the same order, $\sim 10^{-13}$ W m^{-3}. Since cosmic rays may, like starlight, have their origin in nuclear processes in stars, this coincidence is not so surprising. The magnetic field in our Galaxy controls the motion of the cosmic rays and might play a regulatory role in star formation. But the coincidence of these three Galactic energy densities with the energy density of the microwave background, whose spectral shape and isotropy point to a cosmological origin, remains a mystery. Possibly we just have to accept this as a coincidence, as we have to accept the similar apparent sizes of sun and moon.

20. LIFE IN THE UNIVERSE

The antigeocentric viewpoint, which encourages us to believe the cosmological principle, leads naturally to the idea that we are not unique in the universe. Calculations of the probability of other inhabited planets in our Galaxy are rather meaningless at this stage of our knowledge of the origin of life.† But in the framework of the cosmological principle we should assume that there is at least one inhabited planet per galaxy. Naturally it is intriguing to think that somewhere else all the controversies outlined in this epilogue have been resolved. Yet we should not expect to learn these solutions through some intergalactic broadcasting service, for in the framework of evolutionary cosmology, in which galaxies form simultaneously and evolve in parallel, the light from even the nearest comparable galaxy, M31, set off 2 million years ago, long before Andromeda-man would have appeared on Andromeda-earth. We have to solve these problems on our own.

†An estimate that is often quoted is that the number of technical communities in our Galaxy now is equal to 10 per cent of the average survival time of a technical civilization in years.

Further reading

AT THIS LEVEL

Bondi, H. (1960), *Cosmology.* Cambridge University Press.
McVittie, G.C. (1961), *Fact and theory in cosmology*, Eyre &
Spottiswoode, London.
Sciama D.W. (1971), *Modern cosmology*, Cambridge University Press.
Schatzman E. (1974), *Structure of the universe*, Weidenfeld &
Nicolson, London.
Hubble E. (1958), *Realm of the nebulae*, Dover, New York.

MORE ADVANCED

Peebles, P.J.E. (1971), *Physical cosmology*, Princeton University
Press, U.S.A.
McVittie, G.C. (1965), *General relativity and cosmology* (2nd edn).
Chapman & Hall, London.
Weinberg, S. (1972) *Gravitation and cosmology*, Wiley, New York.

REVIEW ARTICLES, AND BOOKS ON SPECIAL TOPICS

Evans, D.S. (ed.) (1973), External galaxies and quasars. *Symp.
int. astr. Un.* 44.
Field, G.B., Arp, H., and Bahcall, J.N. (1973). *The redshift
controversy.* Benjamin, Reading, Massachusetts.
Burbidge, G.R. and Burbidge, E.M. (1967). *Quasi-stellar objects.*
Freeman, San Francisco.
Field, G.B., (1972). Intergalactic matter. *A. Rev. Astron.
Astrophys.* 10, 261.
Longair, M. (ed.) (1975). Confrontation of cosmological models
with observation. *Symp. int. astr. Un.* 63 .
Partridge, R.B. (1969). Primeval fireball. *Am. Scient.* 57, 37.
Rowan-Robinson, M. (1976) Quasars and the cosmological distance
scale, *Nature, Lond.* 262, 97.
Ryle, M. (1968). Counts of radio sources. *A. Rev. Astron. Astrophys.*
6, 249.
Schmidt, M. (1969). Quasi-stellar objects. *A. Rev. Astron.
Astrophys.* 7, 527.
Setti, G. (ed.) (1975) *Structure and evolution of galaxies*,
D. Reidel, Dordrecht, Holland.
Silk, J. (1973) Diffuse X- and γ-radiation. *A. Rev. Astron.
Astrophys.* 11, 299.
Thaddeus, P. (1972). Microwave background. *A. Rev. Astron.
Astrophys.* 10, 305.
Zeldovich, Ya.B. (1965). Survey of modern cosmology. *Adv. Astron.
Astrophys.* 3, 242.

Glossary

---

big-bang models: Expanding universe models in which the density was infinite at a finite time in the past.

black-body radiation: A perfectly efficient radiator or absorber of radiation is called a black-body. When matter and radiation are in complete thermal equilibrium, e.g. during the fireball, the radiation will have a black-body, or Planck, spectrum. The background radiation at microwave frequencies has the spectrum of a 2·7 K black-body and is believed to be a relic of the fireball.

black hole: A region in which the matter has collapsed together to such an extent that light can no longer escape from it.

bremsstrahlung: Radiation from an ionized gas due to electrons moving in the electrostatic field of ions.

Cepheids: Stars whose brightness varies sinusoidally on a period of 2-40 days, the period being directly related to the mean luminosity of the star.

co-moving observer: One who is at rest with respect to the substratum.

Compton scattering: The scattering by free electrons of photons which, in the rest-frame of the electron, have energies greater than the rest-mass energy of the electron ($m_e c^2$). When the scattering results in a boost of the photon energy, it is often called inverse Compton scattering.

Copernican principle: The earth does not occupy a privileged position in the universe.

cosmic rays: Relativistic particles, both nuclei and electrons, continuously bombarding the earth. Some come from the sun, others from pulsars, supernovae, and other violent events.

cosmological principle: The universe as seen by fundamental observers is homogeneous and isotropic.

cosmological time: In a homogeneous universe the proper times of fundamental observers can be synchronized to give a universal, cosmological time.

critical density: That average density of the universe which divides big-bang models that will keep on expanding for ever from those that will ultimately recontract.

deceleration parameter: Measures the rate at which the expansion of the universe is slowing down, in dimensionless form. Must be positive if gravity is the only force acting.

diameter distance: Distance deduced assuming the apparent angular size of an object falls off linearly with distance.

distance modulus: The difference between the apparent and absolute magnitudes of an object, equal to 5 lg(distance in pc) − 5.

epoch of decoupling: The moment in the evolution of a big-bang universe when the matter recombines and becomes transparent to radiation.

field equations: Differential equations which in the general theory of relativity relate the geometry of space-time, described by the metric, to the distribution of matter and other forms of energy.

fireball: The phase in a big-bang universe prior to the epoch of decoupling, when the matter is completely opaque to radiation, and matter and radiation are in thermal equilibrium.

flux: The total energy received from a source per sec per unit area normal to the direction of the source (unit: $W\ m^{-2}$)

flux density: The flux per unit bandwidth (unit: $W\ m^{-2}\ Hz^{-1}$). Also known as the monochromatic flux.

fundamental observer: Observer who is at rest with respect to the substratum i.e. co-moving with it.

Galactic coordinates (b,l): Spherical polar coordinates analogous to latitude and longitude, in which the plane of the Galaxy plays the part of the equator, and the direction of the Galactic centre corresponds to zero longitude.

gravitational wave: According to general relativity waves can be transmitted via the gravitational field, just as light is transmitted via the electromagnetic field.

heavy elements: All elements apart from hydrogen, helium, and the light elements lithium, beryllium, and boron, are referred to as heavy elements by astronomers. On average they make up about 1 per cent of the matter in our Galaxy.

helium flash: The moment in a star's evolution when the core, exhausted of hydrogen, has heated up sufficiently for helium fusion to begin.

Herzsprung–Russell (HR) diagram: Plot of stellar luminosity against surface temperature (or, in practice, colour or spectral type), in which the evolution of stars of different masses may be followed.

homogeneity: A homogeneous universe is one that appears the same to all fundamental observers.

Hubble distance: The distance at which galaxies would have a red-shift of 1 on the basis of the Hubble law. Current value is 6000 Mpc or 2×10^{10} light years.

Hubble law: The red-shift of objects in the universe increases linearly with distance. Most cosmological models predict departures from linearity when the red-shift is no longer $\ll 1$.

Hubble parameter (*or constant*): The slope of the red-shift-distance relation (unit: km s^{-1} Mpc^{-1}). Current value is 50 km s^{-1} Mpc^{-1}.

Hubble time: The time for the universe to double its size expanding at the present rate, if the red-shift is interpreted as due to recession. Current value is 2×10^{10} years.

hydrogen burning: Nuclear fusion of hydrogen into helium in the interior of stars.

H_I *cloud:* Cloud of cool, neutral hydrogen.

H_{II} *region:* Cloud of hot, ionized hydrogen, usually heated by a luminous star.

inertial frame: Frame of reference in which Newton's first law of motion holds.

intensity: The (monochromatic) intensity of an extended source of radiation is the flux (density) per unit solid angle (unit: W m^{-2} sr^{-1} (Hz^{-1})).

interval: Measured in an inertial frame, the interval between two events separated by distance dr and time dt is defined by ds^2 = dt^2 - dr^2/c^2, so that it is zero if the events can be connected by a light signal.

invariant: Scalar quantity that has the same value in all frames of reference.

isotropy: An isotropic universe is one which to a fundamental observer looks the same in every direction on the sky.

Jansky (Jy): Astronomical unit of flux-density: 1 Jy = 10^{-26} W m^{-2} Hz^{-1}.

light year: Distance travelled by light *in vacuo* in one year: 1 light year = $9 \cdot 46 \times 10^{15}$ m.

Local Group: Small group of 20 or so galaxies of which our Galaxy is one of the dominant members (see Table 1; p.4).

luminosity: The bolometric or total luminosity is the energy emitted by a source per unit solid angle per second (unit: $W \, sr^{-1}$). The monochromatic luminosity is the luminosity per unit bandwidth (unit: $W \, sr^{-1} \, Hz^{-1}$).

Luminosity class: Classification of spiral galaxies according to the appearance of their spiral arms, each class having a different mean luminosity. Allows spirals, especially of type Sc, to be used as a distance indicator.

luminosity distance: Distance of a source assuming the inverse square law for radiation holds.

magnitude: Logarithmic scale of brightness used by astronomers, one magnitude corresponding to a change of 0·4 in lg. The apparent magnitude is -o.4 lg (flux) + constant. Photographic (m_{pg}), visual (m_V), photoelectric (U, B, V) magnitudes refer to magnitudes determined with different detectors and filters. Bolometric magnitude is the magnitude that would be obtained if all the light from the source could be detected. Absolute magnitude, the unit of luminosity, is the magnitude a source would have at a distance of 10 pc.

main sequence: Locus in the HR diagram corresponding to the hydrogen burning phase in the life of stars of different masses.

metric: The relationship between the coordinate difference between two nearby events and the interval; defines the geometry of space-time.

neutron star: Cold, degenerate, compact star in which nuclear fuels have been exhausted and pressure support against gravity is provided by the degeneracy pressure of neutrons.

nucleosynthesis: The building up from hydrogen of the elements in the periodic table by means of nuclear reactions, in the fireball for helium, in stars for the other elements.

parallax: The change in apparent direction of a star due to the earth's motion round the sun.

parsec (pc): Distance at which the radius of the earth's orbit subtends one second of arc: 1 pc = 3:26 light years.

peculiar velocity: That part of a galaxy's velocity due to its random motion relative to the substratum.

Planck spectrum: The energy distribution characteristic of a black-body.

principle of equivalence: Gravity may be transformed away locally by choosing a freely falling frame of reference.

proper distance: Distance measured by an observer using radar methods.

proper time: The proper time of an observer is the time measured on a clock at rest with respect to him.

pulsar: Pulsating radio source associated with neutron star.

quasars: Outstandingly luminous quasi-stellar radio sources related to violent events in the nuclei of galaxies.

red giant: Phase in star's evolution after completion of hydrogen burning when outer layers become very extended.

red-shift: Shift in frequency of spectral lines towards the red end of the spectrum, due to recession of source (Doppler shift) or effects of gravity (gravitational red-shift). The cosmological red-shift (see Hubble law) is usually interpreted as implying expansion of the universe.

relativistic: With velocity close to the speed of light.

scale-factor: The function of time by which all distances scale in a universe satisfying the cosmological principle.

Seyfert galaxy: Type of galaxy showing signs of activity in nucleus.

singularity: Point where theory predicts that physical variables, especially the density, become infinite. Most notable is the initial moment of big-bang universes, but also happens to matter inside a black hole.

space-time: In relativity theory the three dimensions of space and one of time are treated as a single four-dimensional space-time continuum.

spectral index: The parameter α if the monochromatic luminosity $P(\nu)$ has the form $P(\nu) \propto \nu^{-\alpha}$, where ν is the frequency.

steady-state cosmology: Model of expanding universe in which all properties are independent of time: continuous creation of matter is necessary to maintain the universe at a constant density.

substratum: The matter in the universe is imagined to be smeared out into a smooth fluid.

synchrotron radiation: Radiation from relativistic electrons spiralling in a magnetic field.

Thompson scattering: Scattering by free electrons of photons which, in the rest-frame of the electron, have energies much less than the rest-mass energy of the electron ($m_e c^2$). No change in the energy of the electron or of the photon results.

white dwarf: Cool, degenerate, compact star, in which nuclear fuels are exhausted and pressure support against gravity is provided by the degeneracy pressure of electrons.

Name Index

Abell G.O., 106, 139
Alfven H., 140
Arp H., 106, 144

Bahcall J.N., 146
Bondi H., 129, 146
Burbidge G.R., 144, 146
Burbidge E.M., 144, 146

Davis R., 11
Dirac P.M., 131, 133, 135

Einstein A., 70, 71, 123
Evans D.S., 146

Fermi E., 9
Field G.B., 146

Gamov G., 135
Gödel K., 137
Gold T., 129

Hawking S., 137
Hoyle F., 129, 131, 131, 143
Hubble E.P., 5, 34, 50-52,
    56, 100, 146

Jansky K., 5
Jordan P., 131

Kellerman K., 143
Kiang T., 139

Longair M., 146
Lundmark K., 50

McCrea W.H., 137
McVittie G.C., 146
Misner C.W., 133, 137

Narlikar J.V., 131, 133

Omnes R., 140

Partridge R.B., 146
Pauli W., 8
Peebles P.J.E., 146

Ryle M., 146

Sandage A., 50, 100, 131, 138

Sciama D.W., 146
Schatzman E., 146
Schmidt M., 146
Silk J., 146

Thaddeus P., 146

de Vaucouleurs G., 106, 134,
    138, 139

Weber J., 12
Weinberg S., 146

Zeldovich Y.B., 133, 146
Zwicky F., 106, 139

Subject Index

absorption lines 14, 15
abundance of elements, 89
Andromeda nebula, 1-4, 114
anisotropic cosmological
  models, 65, 133, 134, 137,
  140
annihilation of matter and
  antimatter, 90, 91, 140
anthropic principle, 140
atmospheric transmission, 17,
  18

background radiation,
  integrated
  19, 22, 93, 109-11, 135
  cosmic microwave, 19, 22,
  57, 63, 86, 87, 93, 110,
  118, 119, 120, 131, 133,
  140-142, 145
  (isotropy of, see isotropy)
  γ-ray, 19, 111, 140, 143
  infrared, 19, 110
  optical, 19, 111, 117
  radio, 19, 110, 120, 139,
  142
  X-ray, 19, 111, 120, 142,
  143
big bang, 22, 28, 57, 69, 78,
  82, 86, 90, 93, 122, 123,
  133, 136, 147
  early stages of(see fireball)
black-body radiation, 12, 13,
  85, 86, 87, 93, 110, 131,
  133, 147
Bologna catalogue of radio
  sources, 106
bounce models, 126, 127, 129
Bremsstrahlung, 120, 147

Cambridge, 3rd catalogue of
  radio sources, 106, 108, 109
Centaurus A radio-galaxy,
  39, 40

Cepheid variable stars, 26, 27,
  48, 51, 147
clusters of galaxies, 5, 22,
  42, 43, 44, 47, 102, 103, 116,
  134, 139, 142
  brightest galaxies in, 48, 100,
  128, 130
  rich, 43, 44, 56, 102, 111, 116
comoving, coordinates, 74, 96
  observer, 63, 147
Compton scattering, 17, 111, 119,
  147
Copernican principle, 64, 65, 76,
  147
cosmic microwave background (see
  background radiation)
cosmic rays, 7, 8, 15, 89, 118,
  144, 145, 147
cosmological, constant, 123-9
  distance scale, 45-8
  model, 19, 63-81, 100, 107
  (see also cosmology)
  principle, 64, 65, 69, 93, 94,
  123, 125, 134, 147
  repulsion, 126
cosmology, big bang (see big
  bang)
  Brans-Dicke, 132, 136, 137
  general relativistic, 74-6, 95,
  96
  Newtonian, 65-70, 74, 75, 93
  special relativistic, 76
  steady state, 107, 129-31, 135,
  137, 138, 150
Crab nebula, 20, 24, 27, 28
creation of matter, continuous,
  129, 130
curvature of space-time 48, 73,
  75, 128
  Gaussian, 130
Cygnus A radio-galaxy, 38, 119
Cygnus X-1 X-ray source, 28, 136

deceleration parameter, 79, 81, 93, 99-102, 105, 107, 128, 129, 148
decoupling, epoch of, 85-7, 118, 141, 148
density of matter, 65, 66, 68, 75, 82, 87, 88, 91, 94, 113-22, 129, 130
critical, 79, 113, 117, 147
density parameter, 128, 129
deuterium, 91
diameter distance, 47, 48, 93, 102-4, 148
distance modulus 47, 148
Doppler shift, 49, 51, 93, 95, 114
dust, 2, 17, 28, 31, 39, 53-5, 84, 117, 118

Eddington magic numbers, 126, 128
Eddington-Lemaitre models, 126, 128
Einstein-de Sitter model, 77, 78, 79, 81, 112, 121
electromagnetic radiation, 6 7, 12
electrons, relativistic, 7, 8, 15-17, 39, 111
emission lines, 14, 15, 40-2, 114
energy, density of radiation, 69, 82, 85, 87, 144
gravitational, 114
kinetic, 114
equivalence, of mass and energy, 71, 82
principle of, 71, 150
Euclidean space, 48, 73, 94
event, 71, 72, 96, 112
evolution of source population, 93, 107, 109, 110, 141
extinction by dust, 53-6, 62

Faraday rotation, 121
field equations, 75, 123, 130, 148
fine-structure constant, 135

fireball, 22, 57, 61, 62, 69, 82-92, 113, 118, 140, 148
flux of radiation, 94, 99, 108, 109, 148
flux-density, 18, 19, 22, 46, 148
frame of reference, 65, 70-3
freely falling, 71, 72
fundamental observer, 63-5, 67, 148
free-free radiation (see Bremsstrahlung)
frequency, 5-7, 17, 18, 20, 21, 49, 50, 93

γ-rays, 6, 18, 19, 21 (see also background radiation)
galaxies, 3, 5, 20, 22, 23, 31-41, 48, 50-2, 84, 100, 101, 112, 134
active, 7, 23, 40-4
counts of, 54-6, 58, 105, 113, 143
elliptical, 3, 33-6, 39, 41, 42, 44, 114, 115
formation of, 113, 116, 121, 122, 133, 137, 138, 141
irregular, 34-6
masses of, 4, 113-16
mass-to-light ratio of, 114, 116
N-type, 40, 41
nuclei of, 35, 39-41, 44
radio-, 20, 23, 38, 39, 44, 101, 103, 104, 107-11
rotation curve of, 114
Seyfert, 23, 40, 42, 111, 144, 150
spiral, 3, 33-6, 38, 40, 113, 115
Galaxy, the 1-4, 12, 15-19, 22, 23, 28-31, 44, 49, 53, 60, 63, 88, 110, 111, 114, 118, 119, 140-2, 144, 145
disc of, 28-31
halo of, 28-30
nucleus of, 23, 28, 31
radio emission from, 16
rotation of, 30, 31, 49, 63
geodesics, 74, 96

globular clusters, 29, 30, 47, 48, 89, 115, 117, 122
gravitational, constant, 131-3, 137
  wave detector, 11, 118
  waves, 9, 12, 118, 148
gravity, 71, 72, 121, 134
  Newtonian, 65, 69, 73, 76

helium, 25, 26, 28, 30, 62, 88-91, 119, 140
  flash, 26, 148
Herzsprung-Russell diagram 25, 27, 30, 148
homogeneity, 64-6, 138, 143 149
horizon, 112
Hubble, constant/parameter, 51, 52, 79, 93, 138, 149
  distance, 50, 84, 149
  law, 50, 51, 67, 99, 149
  sequence, 35, 36, 142
  time, 50, 52, 61, 80, 81, 117, 127, 129, 131, 149
hydrogen, 25, 28, 30, 62, 89, 119
  atomic (H_I, neutral), 15, 31, 36, 114, 119, 120
  burning, 25, 26, 131, 149
  ionized (H_{II}), 120, 121, 130
H_{II} region, 23, 48, 51, 89, 149

inertial frame, 70, 94, 149
infrared radiation, 5, 6, 20, 23, 40 (*see also* background)
instability strip, 26, 27
intensity of radiation, 22, 60, 85, 92, 109, 149
intergalactic gas, 111, 116, 119-21, 130, 142
interstellar gas, 17, 23, 27
interval between events, 72, 96, 149
invariant, 65, 72, 149
inverse Compton scattering (*see* Compton scattering)

inverse square law, 18, 46, 94
isotope, 61
isotropy, 53-7, 64, 67, 74, 133, 137, 138, 140, 149
  of clusters, 56-7
  of galaxies, 53-6
  of microwave background, 57, 64, 93, 131, 133, 137, 145
  of radio-sources, 57

Jeans length, 121

Large Magellanic Cloud, 4, 33
Lemaitre models, 126-8
lepton, 90
light year, 1, 3, 4, 149
Local Group of galaxies, 1-4, 36, 44, 47, 50, 53, 101, 117, 122, 141, 142, 149
Local Supercluster, 58, 138, 139
luminosity, 46, 58-60, 94, 101, 105, 107-09, 131, 150
  bolometric, 12
  class of spiral galaxies, 48, 51, 150
  distance, 46, 48, 93, 98-102, 108, 150
  function of clusters, 48
  monochromatic, 12, 18, 107
  radio, 107
  -volume test, 108
Lyman α, 14, 120, 121

Mach's principle, 132, 137
magnetic field, 16, 39, 111, 118, 121
magnitude, absolute, 4, 46
  apparent, 35, 46, 56, 100, 101, 150
  -red-shift test, 79, 100-02, 128, 130, 131
main sequence, 25, 30, 150
matter-dominated era, 84, 87, 88
meson, 90
Metacluster, 134, 139
metric, 73, 75, 102, 150
  Minkowski 73

Robertson-Walker, 74, 96
steady state, 130
tensor, 73
Michelson-Morley experiment, 71, 141
microwave background (see background radiation)
microwave radiation, 20, 57
Milky Way (see Galaxy, the)
missing matter, 115, 116, 142
molecules, 15, 31, 121

neutrino, 8, 9, 11, 87, 90, 91, 118
telescope, 11
neutron, 91
capture, 26
stars, 7, 24, 27, 43, 150
novae, 27, 28
nuclear reactions, 8, 23, 25, 27, 62, 88
nucleon, 90

Olbers' paradox, 60
Orion nebula, 20, 23, 24

parallax, 46, 48, 50
parsec, 46, 150
peculiar velocity, 50, 63, 150
perihelion advance of planets 74, 132
photon, 7, 8, 13, 17, 73, 84, 90, 91, 96, 99, 118
Planck spectrum, 12, 13, 85-7, 92, 120, 141, 150
planetary nebulae, 27, 47, 89
population, stellar, 28, 31
positron, 91
power-law, 16
pressure, 65, 66, 69, 75, 82, 83
proper distance, 99, 150
proton, 91
protostar, 23, 25
pulsars, 7, 27, 150

quantum gravity, 136

quasars, 7, 14, 17, 23, 40, 42-4, 52, 101, 103, 104, 107-11, 127, 128, 135, 144, 150
local theory for, 117

radar distance, 46, 99
radiation-dominated era, 84, 87, 88
radiation temperature, 13, 85, 96
radio-galaxy (see galaxies)
radio radiation, 5-7, 15-20, 22, 38-41 (see also background radiation)
radioactive dating, 61
radio-sources, 57, 103, 104, 107, 144
counts of, 59, 107, 130, 143
tail, 44, 116
red giant, 25, 26, 150
red-shift, 42, 49-53, 59, 93, 96, 97, 99, 104, 107, 112, 128, 150
cosmological, 51, 61, 67, 101, 127
-distance relation (see Hubble law)
gravitational, 49, 74, 144
intrinsic, 117, 144
Reference Catalogue of bright galaxies, 101, 106
relativity, general, 12, 71-6, 93, 130, 132, 136, 137
special, 70-3, 94, 95
RR Lyrae stars, 26, 27, 48

scale-factor, 58, 74, 85, 123, 150
singularity, 78, 133, 150
de Sitter model, 124-7, 130, 135
source counts, 19, 58, 93-5, 104-08, 110 (see also radio-sources, galaxies)
space-time, 72, 150
curved, 48, 73-75
spectral index, 16, 151
spectral line, 15, 50, 52 (see also absorption lines, emission lines)

spectral type, 25
spectrum, 5.7, 12, 14, 50
standard candle, 38, 46
star formation, 23, 28, 30,
  36, 85, 142
stellar evolution, 25-8, 61,
  89, 111
submillimetre radiation, 6,
  20, 110, 141
substratum, 63, 82, 94, 150
supernovae, 7, 24, 26-8, 48
synchrotron radiation, 15-
  17, 38, 150

thermal equilibrium, 85, 86
thermodynamics, first law
  of, 82
thermonuclear reactions (see
  nuclear reactions)
time, universal cosmical,
  65
tuning-fork classification
  of galaxies (see Hubble
  sequence)

Uhuru catalogue of X-ray
  sources, 44, 106
ultraviolet radiation, 5,
  6, 17, 19, 21
uniformity, 57-9, 109
universe, age of, 60, 61,
  78, 80, 81, 93, 138
  Einstein static, 125-7
  expansion of, 18, 51, 60,
  61, 68, 76, 77, 98, 112,
  113, 124, 125, 129
  hierachical, 134
  open, 112
  oscillating, 78, 80, 124,
  129
  probability of other life
  in, 145
  radiation dominated, 61, 83
  structure of, 45

Virgo cluster, 2, 3, 44, 53,
  58, 101
virial theorem, 114, 116,
  122

wavelength, 5-7, 17, 18, 20,
  21, 49, 93
white dwarf, 26, 43, 150

X-ray radiation, 5-7, 17-19,
  21, 22, 28, 41, 43, 44, 119,
  136 (see also background
  radiation)
  hard, 6, 21, 120
  soft, 6, 17, 21, 119, 120, 142

zodaical light, 19, 21, 22,
  111
zone of avoidance, 54-6

Physical constants and conversion factors

Avogadro constant	L or N_A	6.022×10^{23} mol^{-1}
Bohr magneton	μ_B	9.274×10^{-24} J T^{-1}
Bohr radius	a_0	5.292×10^{-11} m
Boltzmann constant	k	1.381×10^{-23} J K^{-1}
charge of an electron	e	-1.602×10^{-19} C
Compton wavelength of electron	$\lambda_C = h/m_e c$	$= 2.426 \times 10^{-12}$ m
Faraday constant	F	9.649×10^4 C mol^{-1}
fine structure constant	$\alpha = \mu_0 e^2 c/2h$	$= 7.297 \times 10^{-3}$ $(\alpha^{-1} = 137.0)$
gas constant	R	8.314 J K^{-1} mol^{-1}
gravitational constant	G	6.673×10^{-11} N m^2 kg^{-2}
nuclear magneton	μ_N	5.051×10^{-27} J T^{-1}
permeability of a vacuum	μ_0	$4\pi \times 10^{-7}$ H m^{-1} exactly
permittivity of a vacuum	ϵ_0	8.854×10^{-12} F m^{-1} $(1/4\pi\epsilon_0 = 8.988 \times 10^9$ m F^{-1})
Planck constant	h	6.626×10^{-34} J s
(Planck constant)/2π	$\hbar$	1.055×10^{-34} J s $= 6.582 \times 10^{-16}$ eV s
rest mass of electron	m	9.110×10^{-31} kg $= 0.511$ MeV/c^2
rest mass of proton	m_p	1.673×10^{-27} kg $= 938.3$ MeV/c^2
Rydberg constant	$R_\infty = \mu_0^2 m_e e^4 c^3/8h^3$	$= 1.097 \times 10^7$ m^{-1}
speed of light in a vacuum	c	2.998×10^8 m s^{-1}
Stefan–Boltzmann constant	$\sigma = 2\pi^5 k^4/15h^3 c^2$	$= 5.670 \times 10^{-8}$ W m^{-2} K^{-4}
unified atomic mass unit (^{12}C)	u	1.661×10^{-27} kg $= 931.5$ MeV/c^2
wavelength of a 1 eV photon		1.243×10^{-6} m

$1\,\text{Å} = 10^{-10}$ m; $\quad 1$ dyne $= 10^{-5}$ N; $\quad 1$ gauss (G) $= 10^{-4}$ tesla (T);
$0°\text{C} = 273.15$ K; $\quad 1$ curie (Ci) $= 3.7 \times 10^{10}$ s^{-1};
1 J $= 10^7$ erg $= 6.241 \times 10^{18}$ eV; $\quad 1$ eV $= 1.602 \times 10^{-19}$ J; $\quad 1$ cal$_{th} = 4.184$ J;
$\ln 10 = 2.303$; $\quad \ln x = 2.303 \log x$; $\quad e = 2.718$; $\quad \log e = 0.4343$; $\quad \pi = 3.142$